Jump Start Your Career

i n

BioScience

Chandra B. Louise, Ph.D.

Peer Productions
Durham, North Carolina

Jump Start Your Career
in BioScience

Chandra B. Louise, Ph.D.

published by:

Peer Productions
P.O. Box 13724
Research Triangle Park, NC 27709   U.S.A.

Cover design by Boyce McClain

Library of Congress Cataloging-in-Publication Data
Louise, Chandra B.
    Jump start your career in bioscience / by Chandra B. Louise. -- 1st ed.
        p. cm.
        Includes bibliographical references and index.
        ISBN: 0-9661790-0-5
        1. Life sciences--Vocational guidance.
        2. Biology--Vocational guidance.   I. Title.

QH314.L68 1998   570'.23
                                        QBI97-41582

Library of Congress Catalog Card Number: 97-75974

Contents

Contents
(continued)

Preface

This book is the product of my own job seeking adventure. It all began when I started looking for some information on how I could use my science degree to work in the "real world".

I have written this book to include the type of information that I had wanted all in one place, when I was looking for a job. I wrote it with the hope of making it easier for others to find a rewarding career path without having to do all of the research *de novo*.

This book provides you with descriptions of a number of conventional and "alternative" career opportunities, as well as ideas about potential employers including the private sector, government organizations, and academic institutions. You'll also find information about how to identify specific job opportunities and about techniques and resources which can lead you down the path to a scientifically and financially rewarding job.

This book is intended for students, postdoctoral fellows, faculty, career counselors, and anyone else who is interested in learning more about the different types of careers in the biosciences. Although this book focuses mainly on careers in the biomedical and life sciences, you'll find that many of these career descriptions are applicable to other scientific disciplines as well.

If you like this book, please refer your friends to it. Furthermore, if you have any comments or suggestions for improvement, please send them to P.O. Box 13724, Research Triangle Park, NC 27709. Let's continue our scientific tradition of the sharing of knowledge for the good of everybody.

After all, isn't that the purpose of scientific publication?

Acknowledgement

This book is the product of many different experiences from many other scientists.

Special acknowledgements go to Dr. Tom Obrig, Dr. Fred Smith, Dr. Gail Seigel, Dr. Dorothea Sanchez, Dr. Belinda Rowland, Ms. Mary Clark, Dr. Evan Siegel, Dr. Kirsten Vadheim, Ms. Martha Regester, and Dr. Jeroo Kotval for all of their helpfulness with this book. I also thank the numerous scientists, allies, friends, and family members who've provided suggestions, offered help, and encouraged me throughout the writing of this book.

Disclaimer

Scientists live in the world of the unknown. From the beginning of our training, we're taught to question information and to accept nothing as fact. We're encouraged, and often even paid, to gather research data. We're then expected to form our own ideas, hypotheses, theories, and beliefs from this information.

This book is written for scientists. As such, it is assumed that you will critically evaluate the information presented within this book and form your own opinions. *You* are the ultimate source of information upon which you should make your career decisions.

The stories, anecdotes, examples, and other material contained in this book are meant to be for illustrative purposes only. The material contained in this book, except where a particular organization is referred to by name, is *not* intended to refer to any specific person, employer, or other entity. As such, material may have been modified, embellished, and/or otherwise fictitiously created. Furthermore, descriptions of particular organizations are the author's interpretation *only* and are not intended to represent the official mission statements of these organizations. You are referred directly to these organizations for an official mission statement and a description of their activities.

The author has attempted in all parts of this book to be as objective and accurate as possible. By the time you read this, however, some of the information may be outdated, and there may be inadvertent typographical and/or other mistakes in this book. This book is intended to be a useful guide for your job search. Nevertheless, you are expected to verify information before accepting it as fact.

Neither the publisher nor the author shall have liability to any person or entity for any direct or indirect loss or damage caused, or alleged to be caused, by the material contained in this book. If you do not accept the terms of the above agreement, you may return this book to the publisher for a full refund.

Part I
Introduction

Chapter 1
Introduction

You may have just finished your training
Or quit at a job that was draining.
In any event,
Your money is spent,
Your wallet is amply complaining.

From paying that hefty tuition,
You now want to find a position
Which isn't a grind
And uses your mind
And satisfies all your ambition.

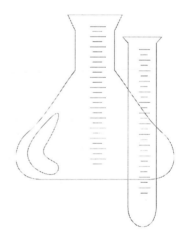

Or maybe you're working already
But looking for something more steady
Or maybe the pay
Is causing dismay
Or maybe your boss is just petty.

Whatever the path that you took,
Whatever the reasons you look,
Careers that are planned
Put you in command,
And that is the point of this book.

Consider the following job interview:

Flexible Fred: "I'm interested in getting a job in your organization."

Potential Employer: "Great. What would you like to do?"

Flexible Fred: "I'll do anything - I'm flexible."

Potential Employer: "So you don't have any position or type of work in mind?"

Flexible Fred: "Not really. I'd like to use my scientific training in some way, but I don't really care exactly what I do. I'll consider anything."

Now consider a contrasting scenario:

Focused Felix: "I am seeking a position in communications at your organization. I believe that my skills could be useful to you."

Potential Employer: "What do you think you can do for us?"

Focused Felix: "Well, in reading about your organization, I noticed that it has a special interest in promoting education about cancer. I have done quite a bit of research in that area. I have also had a lot of experience in relating this information to other people, through both my scientific publications and occasional informational seminars that I have presented to the lay public. I believe that I have a knack for taking complicated scientific problems and relaying them to the public in an understandable way. I've brought a sample of my written work for you to see."

Potential Employer: "That is very interesting. However, we were not thinking of hiring anyone in that capacity at this time."

Focused Felix: "I understand that you do not want to spend the money to hire a person for this type of position. However, I believe that I could bring in a lot of revenue to your organization. For instance, increasing public awareness of your organization would probably increase the number of people who donate money to it. This could bring in a lot more money than you would have to pay me for my salary."

Potential Employer: "You have a point. I'll see what I can do about creating a position for you."

Which kind of interview do you think is more likely to land you a job?

You may think, like Flexible Fred, that a certain degree of flexibility means that you have a wider range of career opportunities available to you. By saying that you're flexible, you figure that you're increasing the number of positions which are available to you.

The problem is, many employers don't see it that way. They interpret a job seeker's "flexibility" as laziness. They think that the job seeker doesn't want to do any work or take any initiative to shape his career. Instead, he is putting the burden on the employer to find him a suitable position within the organization.

Employers generally don't like this. They want to see initiative, enthusiasm, and ambition. Since Flexible Fred has displayed very little of any of these traits, many employers figure that he probably won't display much of these traits on the job, either.

Focused Felix, on the other hand, has displayed a lot of initiative, enthusiasm, and ambition. In fact, he has done all the work for the employer in his job search. He has told the employer exactly how he can help the organization, and exactly what he would perceive his role to be in this situation. He has translated his personal goals into a win-win situation for both himself and the employer.

You, too, can conduct interviews like Focused Felix. You can use this technique to get a job that YOU want and that meets YOUR goals. You just may need a few resources.

Specifically, you need to know two things:

•   You need to know what type of job you want.

•   You need to know how to get it.

This book is designed to help you with both aspects of your job search. Topics addressed in this book are shown in Figure 1.

To many of us, the non-academic world is a mystery. We have no idea how our skills can fit in. Usually, then, we try to seek help from our professors.

Student: "I'm interested in learning about some of the different career opportunities that are available to people with biology degrees. Do you have any advice?"

Faculty Member: "Scientific research is a wonderful career. You have lots of room for creativity and freedom. You also don't have to keep the strict hours of the business world. You work very hard, but it's on your own schedule. You should consider it."

Student: "What about doing research in a pharmaceutical company? What do you know about it, and do you think it is a good career path?"

Figure 1

Topics Addressed in This Book

•Descriptions of specific positions for scientists

•Descriptions of potential employers of scientists

•Hints on how to identify specific job opportunities

•A discussion of the educational requirements for various positions

•A listing of resources likely to be helpful to you.

Faculty Member: "Well, the major disadvantage to doing research in a company is that you cannot work on what you want to. The company can switch you off a project at a moment's notice. Sometimes you may have invested lots of work into the project. That can be very frustrating. You don't really want to do this, do you? You are so good at independent research that it would be a shame to see you sell your creativity short. Besides, companies often don't let you publish your work, which takes the fun out of it, and they are too focused on the commercial value rather than on the science."

If you're like me, you've gotten several things out of this conversation:

•   You know that there are at least two career paths available to you. Both involve research, but for different purposes.

•   Of the two research environments, academia is better.

Based on this advice, it's no great wonder why the academic research environment is getting so overcrowded!

It takes a while before it occurs to many of us that faculty, just like us, have not been out in the "real world". They have not been trained as career counselors. They simply don't know what's out there. Like everyone else, they are also opinionated. They truly believe, based on the realm of their own experience, that the academic research environment is the best place to work. Let's hope so - after all, it is the career path they've chosen!

So where do you go from here? If you're lucky, you may have contacts who work at other jobs. If so, you can try approaching these people to get a better feeling for the types of careers which exist in their line of work.

Student: "I am interested in careers in scientific research in a pharmaceutical company. What do you know about it, and do you think it is a good career path?"

Pharmaceutical Researcher: "I really like working in a company. I don't have to compete for funding. I also like the applied nature of the research - after all, science is a business. What's the point in doing research that isn't going to help anyone? I also like the sense of solidarity that comes from working within a company. We're all working toward the same goals. It seem that all too often, academic labs are competing with one another rather than helping each other."

You may get a different opinion of the type of work than a faculty member may give you, and sometimes you can learn about new opportunities this way. This is in fact a good way of researching the different types of careers that are available to you. By collecting a number of different opinions from a number of different people, you can start to get a good feeling for what types of careers you may wish to pursue.

However, this method is rather slow and tedious if you are trying to learn about a large breadth of career opportunities which are available to you. Besides, this method requires some knowledge. You need to know what to ask, and of whom. For this reason, many people find themselves in a vicious circle of ignorance. This problem is compounded by the fact that many working people are very busy and don't have much time to explain the workings of the world to job seekers.

This book is designed to take the mystery out of the working world and how it functions. The ultimate goal is so that you too can conduct a job interview like Focused Felix above. Focused Felix knows what he wants, and he states exactly how he will also contribute to the organization with which he is interviewing. He may not be the applicant with the highest grade-point average, or the one with the most publications. However, he has set himself apart from the rest of the applicant pool by knowing his job market and by demonstrating how he may be able to contribute to the organization to which he is applying.

This book focuses mainly on careers in the biomedical and life sciences, although you'll find that many of these career descriptions are applicable to other scientific disciplines as well.

Please note that this book does not cover general skills, such as resume preparation and interviewing skills. For this type of information, the reader is referred to other books, such as <u>To Boldly Go</u> by Peter Fiske (written by a scientist for scientists), and <u>What Color Is Your Parachute</u> by Richard Nelson Bolles. There are also plenty of other good books on these general topics, and you are urged to refer to one of them.

It is important to understand that some of the careers described in this book may not be quite what you had envisioned when you first entered the scientific community. That's okay. Many scientists who have pursued both the "conventional" and the so-called "alternative" careers have found that they enjoy their jobs much more than they would've ever imagined. So keep an open mind - it will serve you well in science as well as in your job search.

Good luck in your adventure!

Part II
The Opportunities

Chapter 2
The Pharmaceutical Industry

The purpose of these corporations
Is making new drug formulations
They do all the science
Within full compliance
Of guidelines and strict regulations.

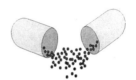

Bob learned about the pharmaceutical industry during one of the most important days of his life. It was the morning that he was to defend his Ph.D. thesis, and he was sick.

Knowing that rescheduling his defense date could be difficult, he took a trip to the pharmacy. He approached the pharmacist:

"I'm looking for some medications to help me. I really have to function today, and I don't feel well. Normally, I don't take drugs, but I'm wondering if you could suggest some medications that might help me make it through the day."

"What are your symptoms?", the pharmacist asked.

"I'm congested, and I have chills and muscle aches. I think I probably also have a fever."

"There are medications which can help you with your congestion, and also some which can help you with your fever. Of course, you want to have medications which won't make you sleepy, and ones that can be taken together. Let's see what we have."

The pharmacist suggested several medications to Bob.

Bob thought for a minute. "How do you know that these medications can be taken together, and that they won't make me sleepy?", he asked. "Are these the results of controlled clinical trials, or just the reported experience of people who just happen to have taken these medications?" It occurred to Bob that unlike his experiments in the lab, human experiments were probably harder to control, and more difficult to both get accurate data and to interpret. That is, if these experiments were conducted at all. Maybe these results were just things that the pharmacist had heard from people who came into the pharmacy. Maybe they were not supported by any real scientific data at all!

"Oh, no", the pharmacist replied. "These findings are from controlled clinical trials. Federal law requires that these claims be supported by actual clinical trial data. Potential product hazards must be written into the product label. These products undergo extensive preclinical and clinical testing."

For the first time, Bob really gave some thought to the types of work conducted by pharmaceutical companies. Surely, pharmaceutical companies must employ lots of scientists in lots of different capacities. Since Bob would be looking for his first "real" job soon, this thought was comforting to him.

The following information is designed to help you to understand the different types of career opportunities that are available in the pharmaceutical industry.

The Pharmaceutical Development Process

To fully understand the career opportunities that exist within the pharmaceutical industry, a basic understanding of the pharmaceutical development processes is very helpful. Thus, a quick overview of the pharmaceutical research, development, manufacturing, and marketing process is shown in Figure 2. To provide you with a more thorough context for the jobs within the overall development process, a more detailed explanation of pharmaceutical development is presented later in the chapter.

Positions within the Pharmaceutical Industry

This section is designed to give you an idea of the different types of biologically- and medically-oriented jobs which exist within the pharmaceutical industry. Please note that jobs within the industry can be rather specialized. That is, pharmaceutical companies can have *entire departments* of people who perform only one or several portions of the overall pharmaceutical development process. **This means that you can think of *every task* as a potentially separate career direction!**

You'll find it helpful to keep this idea in mind when reading through the rest of this chapter, and indeed, the rest of this book. For a more detailed explanation of this concept, please see Chapter 9.

Each step of the pharmaceutical development process involves a different set of tasks. For instance, clinical research scientists do not usually work in the laboratory, whereas basic research scientists often do. For a more complete description of the entire development process, including clinical research, please refer to the latter part of this chapter, where it is described in much more detail.

In the meantime, some examples of biologically- and medically-oriented jobs within the pharmaceutical industry are shown in Figure 3.

Now, let's go into a little more detail about each of these positions.

Basic and Applied Research Scientists

Research scientists are employed in preclinical, clinical, and

Figure 2
The Pharmaceutical Development Process - A Quick Overview

Steps in the Pharmaceutical Development Process

• **Basic research and discovery**

• **Observations of the agent in cell culture and animals**
(absorption, distribution, metabolism, excretion, pharmacology, toxicology, immunogenicity, etc.)

• **Clinical studies in humans**
-Phase I (healthy volunteers)
-Phase II (small numbers of "diseased" patients)
-Phase III (larger numbers of "diseased" patients)

• **Submission of a New Drug Application**
(for approval to sell the drug in the U.S. Separate applications may also be submitted to foreign countries)

• **Product sales and marketing**
-Phase IV clinical trials (to support additional claims about the product, or to expand the number or type of indications for which it is being used)

Other Important Processes:

• **Product Manufacture**
(cost of manufacture, product stability, product formulation and delivery vehicle, ability to produce the product on a large scale, consistently and according to regulations)

• **Determination of the Potential Product Market**
(How large is the market? How much is the market willing to pay for the product? What competition does the product have from other products or treatments for the same indication? Is it cost-effective to develop the product in relation to potential revenues from sales of the product?)

manufacturing aspects of the pharmaceutical development process. Research scientists discover new compounds, design research experiments, collect and interpret preclinical and clinical data, and figure out the best way to manufacture and test the quality of pharmaceutical compounds.

Basic research and discovery usually involves bench-level research. In fact, it can be a lot like academic research, except that some projects may be more proprietary in the pharmaceutical industry than in academia.

"Applied research" is like basic research and discovery except that, as its name suggests, it is being conducted for a very specific purpose, i.e., for some very definite use (whereas basic research and discovery can be more "exploratory" in nature). For example, a team of scientists may work on development of methods for drug manufacture. Another team of scientists may

Figure 3
Some Careers in the Pharmaceutical Industry

Pharmaceutical Career Areas

- Basic and Applied Research Science
- Medical Monitoring
- Drug Safety Surveillance
- Data Management
- Scientific Writing
- Manufacturing Production
- Auditors and Inspectors
- Clinical Laboratory Testing
- Product Sales
- Technology Transfer
- Policy Analysts
- Information Services
- Financial Advising

- Clinical Research Study Monitoring
- Computer Programming
- Pharmacoepidemiology
- Biostatistics
- Regulatory Affairs
- Quality Control
- Quality Assurance
- Management and Administration
- Product Marketing
- Patent and Regulatory Law
- Technical Training
- Product Purchasing
- Computer Support

work on the development of drug delivery systems designed to transport the drug effectively to its target cells and/or organs within the body. Yet another team of scientists may work on optimizing the chemical composition of the agent with regard to certain properties (such as pharmacology, toxicology, immunogenicity, and even novelty, i.e., patentability).

Clinical Research

Clinical research involves the design, monitoring, and interpretation of clinical trials and clinical development programs (the data of which usually comes from clinicians such as physicians in research clinics). Clinical research usually does not involve benchwork, unless you are working in a clinical laboratory. For a more thorough description of the clinical research process, please see the overview of the pharmaceutical development process provided later in this chapter.

Clinical Research Study Monitors

Clinical research study monitors (often called clinical research associates or clinical *research scientists) are representatives of the pharmaceutical company who maintain contact with clinical investigator sites during the conduct of a clinical trial. These individuals are responsible for training these sites and ensuring that the sites are conducting the sponsor's clinical research studies according to the study protocol and according to good clinical practices. Clinical research study monitors contact their sites regularly and visit them periodically to review their research facilities and case report forms for the patients who are enrolled in the study. They are available to answer study-specific questions from their sites, and they keep track of patient recruitment and enrollment to make sure that the study is going along according to the schedule outlined to them by the pharmaceutical company.

Clinical research study monitors may also help to develop protocols and case report forms for new clinical studies. In some companies, monitors may also play a role in the analysis of data and the writing of clinical study reports.

Medical Monitors for Clinical Studies

Clinical studies are conducted in real human beings with real medical issues. Thus, pharmaceutical companies have medical monitors, who are usually medical doctors, who are available to answer complicated medical and clinical study protocol-related questions and to serve as a resource to study staff at investigator sites. Medical monitors may also review serious adverse events to decide whether they are unexpected or require expedited reporting to the Food and Drug Administration (FDA).

Pharmaceutical Research Clinics

Phase I studies are conducted in healthy volunteers in order to determine the absorption, distribution, metabolism, and excretion of the study drug. Thus, these studies involve the collection of numerous blood and urine samples, as well as other clinical observations of the research subjects. Phase I units can be a source of employment for medically trained staff who are able to observe subjects, collect samples from them, and treat any adverse reactions or events that may develop in these subjects.

Most clinical studies in Phases II-IV are too large to be conducted in any single clinic and are usually conducted in conjunction with one or more outside investigator sites (for additional information, please see their description provided later in this chapter).

Drug Safety Surveillance and Pharmacoepidemiology

Pharmaceutical companies must keep track of, and report, certain "serious" adverse events which happen during their human clinical trials. This is usually one function of a drug safety surveillance unit. Adverse events which are considered to be serious, unexpected (i.e., not previously reported to FDA), and related to study medication must be reported to FDA within a certain specified time period after a pharmaceutical company learns of such an event.

In addition to expedited reporting of certain serious adverse events, pharmaceutical companies try to determine causal relationships between pharmaceutical agents and the development of certain adverse events or other conditions or diseases. This process is known as "pharmacoepidemiology".

Epidemiologists

In addition to pharmacoepidemiologists, pharmaceutical companies often have epidemiologists who study the etiology and prevalence of diseases and conditions of interest to the pharmaceutical company. For instance, a pharmaceutical company may want to determine the potential market size for a particular type of therapeutic agent. Perhaps it wants to better understand the etiology, mechanisms, and pathogenesis of certain types of diseases for which a new or improved pharmaceutical agent may be useful.

Epidemiologists can be bench-level scientists who work on the mechanisms of disease, and/or they can be scientists who look at the development of disease from a more clinical or statistical level. They may be considered part of a marketing group, a drug safety surveillance group, a basic scientific group, or they can work in any other number of areas within a pharmaceutical company.

Data Management

Pharmaceutical companies develop databases for many different purposes including the analysis and reporting of the results of preclinical studies, pharmacology and toxicology studies, clinical studies, and epidemiological studies. In each case, the company needs individuals who are able to generate a meaningful database which can then be analyzed and presented in a useful way. Sometimes this involves manipulation of the data, such as the "coding" of adverse events and concomitant medications into generic names or preferred terms. This type of data manipulation is intended to allow the data to be more easily grouped, understood, and/or analyzed.

Biostatisticians

Biostatisticians can provide valuable input to pharmaceutical researchers on issues such as sample size and statistical analysis of study endpoints. Statisticians work in conjunction with other scientists to design and analyze the study endpoints. For randomized studies, statisticians also produce the randomization code.

Statistical input is particularly important for studies in epidemiology, preclinical and clinical pharmacology and toxicology, for all phases of clinical research, and for other types of studies which involve comparisons between different study groups and which have defined study endpoints. Biostatistical input both at the beginning and the end of a study can be tremendously useful in creating and conducting a meaningful and statistically valid study.

Computer Programming

Electronic data analysis, like other types of data analysis and manipulation, requires the ability to compare data points, provide descriptive statistics, tables, and graphs, and to perform statistical testing on the data. These analyses are often custom-made for a particular research study or purpose and therefore require computer programming to generate the desired output. Many companies have a separate group of individuals who perform this programming, with input from scientists and biostatisticians. In some companies, biostatisticians or scientists may perform this programming themselves. In any event, it is often helpful for a computer programmer to have some scientific knowledge, so that he or she can better understand and accommodate the analyses being performed.

Medical, Regulatory, and Other Types of Scientific Writing

Pharmaceutical companies produce many different types of documents for many different purposes. These can include publications in scientific journals, documents for regulatory submissions (such as study reports, IND, and NDA applications), documents for physician information (such as product monographs and clinical investigator brochures for experimental products), and documents for both patients and their physicians (such as the drug package insert).

In some companies, these documents may be produced by individuals who also have other roles in the company (for example, as research scientists). However, many companies have separate writing departments which produce some or all of these different types of documents. Companies with separate writing departments employ full-time writers who interact with the scientists, statisticians, and regulatory personnel to produce these final documents. Medical writers may interact with individuals from other departments (such as marketing) as well.

Regulatory Affairs

The pharmaceutical industry is a highly regulated industry. For instance, the manufacture of agents, the conduct of clinical trials, and marketing of pharmaceutical products are all regulated by the Food and Drug Administration (FDA).   Manufacturing plants are also regulated by the Environmental Protection Agency (EPA). All workplaces are regulated by the Occupational Safety and Health Administration (OSHA) of the federal government. Pharmaceutical companies may also be regulated by state and local regulatory authorities.

Someone must make sure that the pharmaceutical company is operating in compliance with all of these different regulations. Oftentimes, this role is played by a team of individuals who specialize in regulatory affairs. These individuals can serve as a resource to other departments (such as the clinical research and marketing departments), to ensure that they are in compliance with international, federal, state, and local regulations. Individuals in regulatory affairs may review documents, answer questions, and/or participate in the development of company processes from a regulatory standpoint.

Manufacturing Production

For pharmaceutical companies with their own manufacturing plants, individuals with scientific backgrounds are needed to run the day-to-day operations of these manufacturing facilities. For instance, technicians are needed to perform the techniques of manufacture, and scientists and managers are needed to oversee the entire process. Certain individuals perform quality control and quality assurance measures on intermediate and final products. Others "validate" manufacturing techniques to show that they yield consistent and reproducible results. These processes of quality control and validation are also discussed below.

Quality Control and Quality Assurance

"Quality control" is the routine examination of products for quality. For example, a manufactured product may be tested for purity or freedom from microbial contamination. Research reports may be checked against statistical output to ensure that there are no mistakes in reporting. Thus, quality control involves the inspection or testing of products on a day-to-day basis, in order to catch any mistakes or problems with an already-manufactured product.

"Quality assurance" is the process of implementing policies and procedures to minimize the chances that a particular mistake will happen in the future. The creation of company standard operating procedures (SOPs) is one such example. SOPs are designed to ensure that company tasks are performed homogeneously and in a manner which yields a reliable and high-quality product. Quality assurance may also involve inspections of files, facilities, or work, in order to identify potential quality problems and to suggest improvements and/or solutions for these potential problems, before they become actual problems. In this way, quality assurance programs are implemented before-the-fact, as a preventive measure.

Both quality control and quality assurance measures are designed to ensure that a pharmaceutical company produces a uniform, reliable, and high-quality product. This is important from both a regulatory and a marketing standpoint.

"Validation" is the process of demonstrating that a technique or process works consistently, to produce the desired end result. For instance, a manufacturing process (or a step in the process) can be validated to show the reliability of a particular reaction in producing a particular type of end-product. If a company can demonstrate sufficient consistency and reliability of a particular process, it may be able to eliminate or reduce some of the quality control measures that would otherwise be needed.

Other Auditors and Inspectors

Pharmaceutical companies conduct audits and inspections for a wide variety of purposes in addition to assuring the quality of their products, and the processes leading to the generation of the products. These include assurance of compliance to other types of regulations, guidelines, or policies. For instance, industrial hygienists and health physicists may inspect workplace facilities to ensure worker safety. Individuals with knowledge of environmental regulations may inspect manufacturing facilities to ensure compliance of these facilities to the regulations of the Environmental Protection Agency (EPA). Obviously, the types and scope of inspections conducted by companies depend on company size, focus, and philosophy, as well as on governmental regulations.

Clinical Laboratories

Some pharmaceutical companies have their own clinical laboratories which perform tests on preclinical and clinical laboratory samples which are collected during a trial. These laboratories require medical technicians to perform techniques, as well as scientists to direct the laboratories. These tests can include analyses of simple hematology and blood chemistry parameters, or they can include assays for a particular type of hormone, antibody, or even the pharmaceutical compound itself. Even pharmaceutical companies which do not have their own preclinical and/or clinical laboratories often have individuals who can supervise and provide direction to an outside preclinical and/or clinical laboratory (which is contracted to do the work for the pharmaceutical company).

Management and Administration

Pharmaceutical companies, like all organizations, have individuals who decide on the overall directions of the company, both scientifically and from a business perspective. These individuals may make decisions regarding individual scientific studies, as well as overall programs of the company. Additionally, pharmaceutical companies need individuals who can manage projects, programs, budgets, grants, and personnel on a day-to-day basis, and from an administrative as well as a scientific standpoint. Companies who outsource certain tasks need to have individuals who can liaise with the contractors to ensure that the work is being performed according to the needs and budget of the company.

Managers who organize and coordinate individuals within any one particular functional group (for example, within a particular department) are often called "line managers". Managers who coordinate particular projects across functional groups are often called "project managers". Individuals who coordinate entire development programs (preclinical, clinical, and/or marketing) for a particular product are often called "program managers" or "product managers". Companies also have financial managers who keep track of finances and budgets within the company.

There are also individuals who make decisions regarding the overall scientific direction of the company or of a particular program. These individuals are often called "chief scientific officers" of the company. Other chief officers and vice presidents also participate in the overall decisionmaking process, each bringing his or her own perspective to the table (e.g., financial, regulatory, scientific, medical, and so on).

Product Marketing

Marketing departments study the markets for the company's products, including the potential market size, competing products or alternative treatments for the same indication, and the price the market is willing to pay for a pharmaceutical product. This information then can be used for development of product labeling, as well as product advertising campaigns.

Marketing departments also look for ways to expand the market or to increase sales by making specific claims about pharmaceutical products (for instance, by showing that the company's products are safer, more efficacious, or provide a better quality of life than do competing products or treatments). The FDA requires that these types of claims be supported by data from clinical trials. Thus, pharmaceutical marketing departments may design clinical trials to test these claims and to support additional publications and submissions to the FDA for new or improved indications.

Product Sales

Once a company has decided on the markets for which a product is intended, the company then sends representatives out to "detail" the product to the physicians who are likely to prescribe it. These sales representatives, who are often referred to as "detail" people, explain the agent, its intended indications, and its benefits to these physicians, with the hope of increasing the sales of the product. These sales representatives often travel to hospitals, clinics, conventions, and physicians' offices to meet with physicians face-to-face and show them the company's products.

Technology Transfer/Product Licensing

Sometimes pharmaceutical companies want to "in-license" a patented product, technology, or other invention, in order to develop it for a particular use or indication, or to use it for an assay, a manufacturing technique, or for some other purpose. Thus, pharmaceutical companies may license any number of products or inventions including pharmaceutical agents, techniques, procedures, scientific software, or any other type of invention or product. Technology transfer involves a granting of a legal right for an organization or individual to use another's patented product, invention, or idea.

In-licensing is quite frequently the method by which "virtual" companies acquire their pharmaceutical products. "Virtual" companies are discussed later in this chapter and throughout this book. Additionally, in-licensing can be used by larger pharmaceutical companies to acquire additional products or technologies which were developed by others but in which they are interested.

Many organizations have individuals who specialize in identification of potential markets for the product or invention which the company would like to in-license or out-license. These individuals can also execute the license agreement.

In-licensing involves the purchase of the rights to use a product or invention from another company or individual. Out-licensing involves the sale of those rights to another company or individual.

Law

Pharmaceutical companies have a need for lawyers who understand scientific and technical issues as well as law. This need is particularly evident in the area of patent law, in order to determine whether a product or invention is actually worthy of a patent. That is, patent lawyers need to be able to determine whether a product is novel and whether it has potential to be useful. A good technical background can also help a patent lawyer to foresee other potential scientific or legal issues which may arise (for instance, uses of the product for other indications or modification of its molecular structure) for which the company may wish to protect itself.

Technical knowledge, as well as legal knowledge, can be useful in other areas of law as well. For example, contract law can be important as it pertains to contracts between companies and contract research organizations (see Chapter 7), or as it pertains to contracts involving technology transfer. Environmental law and bioethics may be important interests of a pharmaceutical company as well.

Other Support

In general, a wisely managed pharmaceutical company will provide support for their employees in any way which they believe will increase the company's overall bottom line. For instance, the standardization of a procedure (or a set of procedures) may allow the company to get a discount on reagents, to coordinate the efforts of multiple groups of individuals, and/or to develop a centralized system for training new employees in the procedure. In this case, a company may create one or more positions for individuals who select and purchase the equipment and/or reagents. These individuals may also train employees in the procedure or in the use of the equipment and/or reagents they have purchased.

Figure 4

Support Positions within the Pharmaceutical Industry

Some Support Positions

- Policy Analysts
- Government Lobbyists
- Librarians /
  Information Services
- Financial Advisers

- Technical Trainers
- Product Buyers
- Technical Recruiters /
  Human Resources
- Computer Support People

There are many examples of scientific support positions which may benefit pharmaceutical companies. For instance, many companies maintain technical libraries which offer services such as literature searches, Internet access, or other useful materials (such as access to bioinformatics software). Some companies have technical recruiters and/or human resources representatives who have enough scientific know-how to identify qualified individuals for certain types of technical or scientific positions.

Pharmaceutical companies may also hire individuals whose job it is to keep track of important scientific policy issues which could adversely or positively affect the industry. Pharmaceutical companies, like many other industries, can be profoundly influenced by governmental and public policies and regulations. An awareness of these issues can help a company to lobby either for or against certain types of changes, and to be prepared for them in the event that they happen. Since many of the individual workers within the company generally do not have time to track all of these policies and how they affect the industry, policy analysts can be employed by the companies to fulfill this role. Additionally, companies may employ lobbyists or other company representatives who try to promote or prevent changes in policies.

Publicly owned companies may have individuals who keep track of how company strategies could affect stock prices. Companies may have other types of technically-oriented financial advisers as well.

Of course, the availability of these types of positions depends on the company's scientific focus, financial situation, size, and philosophy, among other factors. Nevertheless, the possibilities for these types of positions are far-reaching. With enough creativity and persuasive power, you may even be able to talk a company into creating a position for you!

Now let's shift gears and give you a context for these positions within the pharmaceutical development process!

Overview of the development process for pharmaceutical and biological agents

You may now be wondering how these jobs fit in to the overall operation of a pharmaceutical company. The following section provides you with a context for these positions within the overall pharmaceutical development process. The purpose of this section is to describe the nature of the jobs listed above, and where to find them within the pharmaceutical industry.

First Step - Research and Discovery

Basic research and discovery are usually the first steps leading to the development of a pharmaceutical agent with commercial value. This work often resembles the basic scientific research which occurs in academic laboratories. It can involve any number of different scientific disciplines. Scientists in this type of work look for diseases to prevent and/or treat, and compounds which possess that type of activity.

The jobs within basic research and discovery, as well as in clinical research, are usually very much like other types of bench research. Of course, some research is more applied, whereas other research is more basic in nature. Some research lends itself to considerable autonomy, whereas other projects require a close coordination with others. Nonetheless, the research is usually bench-level, although the data gathering and analysis may involve statistics, computers, and relatively sophisticated computer databases and analysis programs (which can lead to jobs in these areas as well).

Second Step - Preclinical Research and Development

Once a scientific research and discovery team discovers
that a particular compound may have commercial value for the
treatment or prevention of a particular condition or disease entity, the company
may then decide whether to proceed to the next step. This is the testing of the
compound in cell culture and in animals. The purpose of this research is to study
the toxicological properties of the compound, as well as its absorption,
distribution, metabolism, excretion, and pharmacological properties within
whole animals. These results are used to determine whether the agent is likely to
be safe and effective in humans, and at what dosages.

If there is a good animal model for the indication for which the agent is
being developed, further research is done in order to determine the efficacy of the
compound. That is, the company wants to know whether the agent works as
intended in the animal model.

Of course, if the compound has already been used in humans for another
indication at similar doses, some (or all) of this work may not need to be repeated.
However, if the manufacture or formulation of the compound is different, then
some tests will need to be performed.

In any event, this stage of pharmaceutical development is generally
called "preclinical research", since its purpose is to gather the necessary data to
justify introducing the agent into humans.

The Investigational New Drug Application

The next step in the overall pharmaceutical development process is to
test the new compound in humans. When a pharmaceutical compound reaches
this stage of development, the tasks are now considerably different than in basic
scientific research.

Before the company can take this next step, though, it has a few
regulatory hurdles to overcome. The United States government does not allow
companies to simply introduce the compound into humans without prior approval
from the government. Suppose the agent is not safe, or suppose it doesn't work?
Certainly there is no foolproof way of knowing this until the agent is in fact tested
in humans. However, the government wants to eliminate, at least where possible,
the needless testing of compounds in humans which are likely to be unsafe or to
be of no benefit to them. This is one of the purposes of the U.S. Food and Drug
Administration (FDA).

The FDA has guidelines and regulations which govern the testing of
pharmaceutical agents. These rules specify that a company must submit an
Investigational New Drug Application (IND) to the FDA before first testing a
compound in humans. This IND application summarizes the results of preclinical
testing of the compound. The information presented in this application should be
sufficient to allow the FDA to decide whether to allow testing of the compound
in humans.

Preclinical researchers may prepare the IND application themselves, or
the application may be prepared by a separate group of writers. Decisions

regarding the preparation and submission of a IND application typically also involve the pharmaceutical company's regulatory and marketing departments and upper management. Others may become involved as well.

Third Step - Clinical Research and Development

Once the pharmaceutical company gains FDA approval to begin clinical trials in humans, the "clinical development" stage begins. Clinical research and development generally occurs in "phases", each of which has its own purpose.

Phase I of Clinical Trials

Phase I studies are conducted in order to obtain information about how the drug is absorbed, distributed, metabolized, and excreted by the human body. These parameters are referred to as the pharmacokinetic properties of the drug. Phase I studies are also designed to test the safety of the drug in humans, and the effect of the drug on the human body. The effect of the agent on the body is called the pharmacodynamic properties of the drug. Phase I studies are used to determine the maximum tolerated dose, the optimal dose, and effective routes of administration of the drug. Phase I studies are conducted in healthy volunteers.

Phase II of Clinical Trials

Phase II studies are pilot studies which are designed to test both the safety and efficacy of the drug in small numbers of "diseased" patients. These studies utilize a wide range of doses, in order to determine the optimal dose or dose range for treatment of the condition. These studies are the prelude to the larger Phase III studies, which test both the safety and effectiveness of the drug.

Phase III of Clinical Trials

Phase III studies are designed to test the safety and efficacy of the drug in large numbers of "diseased" patients. The results of Phase III studies can be used to form the basis of a New Drug Application to the Food and Drug Administration (see below). Thus, these studies need to have sufficient numbers of patients to achieve statistical power and significance, as well as a diversity of patients within the population, for example, patients who are taking different types of concomitant medications and/or patients with other underlying conditions.

A Note on the Conduct of Clinical Trials

Clinical trials are conducted in humans. This has several important implications with regard to the types of positions in clinical research:

> • All clinical trials are conducted under the auspices of a licensed medical doctor (M.D.), or some other individual who is licensed to take care of patients.

• To protect the safety and best interests of the research subjects (and of the eventual consumer of the drug), clinical trials of pharmaceutical agents are tightly regulated by the U.S. Food and Drug Administration (FDA). Thus, both investigators and pharmaceutical companies must know and comply with the regulations of the FDA.

• Many clinical studies, particularly those in Phase II through Phase IV, need to be large, thereby requiring access to large numbers of patients who have the disease or condition which is being studied. Because of these requirements, clinical studies are often conducted over a large geographic area and at a large number of investigational "sites". This is somewhat less of a problem for Phase I studies, which are generally smaller and conducted in healthy volunteers.

In order to understand the types of job opportunities which are available in clinical research, it is instructive to go over the basic steps which are involved in conducting a clinical trial. For the purposes of this discussion, we will assume that the Investigational New Drug Application (IND) has already been filed with the FDA.

Figure 5
Basic Steps of a Clinical Trial

---

1. The pharmaceutical company decides to "sponsor" (i.e., pay for) a particular type of clinical trial.

2. The study protocol is written and the case report forms are developed. The purpose of these two materials is to document specifically how the study is to be conducted.

• The study protocol contains information such as the types of patients who are to be included and excluded from the study, the procedures that are to be done at each visit of the patient to the clinic (and how these procedures are to be done), and other study-specific information such as what to do in the event that a patient discontinues or has a serious adverse event during the study.

• The case report forms contain blanks and boxes that are to be filled in by the investigator in order to capture the data of the study and the important information which is needed to analyze the results of the clinical trial. Each case report form contains the complete set of data for a single patient.

(continued next page)

Figure 5 (continued):

3. Appropriate clinical investigator sites (which are clinics usually supervised by licensed M.D.s or by some other licensed medical professional) are chosen for the study. Contracts are set up between the sponsoring pharmaceutical company and these investigator sites. Sites are usually chosen by pharmaceutical companies based on criteria such as their ability to recruit and enroll appropriate patients for the study, their prior research experience, their medical specialty, and their geographic location. The pharmaceutical company will usually visit the sites prior to final selection, to view their facilities and discuss the study with them.

4. Investigators are trained in the conduct of the study by the pharmaceutical company. Regulatory documentation for these sites is collected and filed with the FDA. Sites are then "initiated" by a representative of the pharmaceutical company, and investigational drug is shipped to the sites.

5. Sites begin to recruit and enroll patients into the study. As the study progresses, a representative from the pharmaceutical company (also called a study monitor or clinical research associate) calls the site regularly to discuss the site's progress in the study. The representative serves as a resource to the site throughout the study to answer any questions and to address any issues which may arise. The representative also visits the site occasionally to view the site's facilities, paperwork, and case report forms. This is done in order to assure compliance to the study protocol and to federal regulations.

6. During the conduct of the study, several things may happen:

• Serious adverse events may occur in study subjects. Serious adverse events, as their name implies, are medical occurrences such as the development of cancer, any condition which requires hospitalization, or other criteria which are defined by the FDA which can occur in study participants and which may or may not be related to study drug. All serious adverse events, regardless of relationship to study drug, are reported to the sponsoring pharmaceutical company from the sites. Additionally, certain serious adverse events, such as those which are unexpected and/or possibly related to the investigational drug, must be reported to the FDA within a specified time frame and according to a specific format. The sponsoring company usually does the reporting to the FDA.

• Laboratory tests are performed during the study for the purposes of gathering safety information. Laboratory tests are also used, when

(continued next page)

Figure 5 (continued):

appropriate, to determine drug efficacy, immunogenicity, or other information as well. In many of these cases, laboratory samples are collected by the physician and are sent to a clinical laboratory for analysis of the samples, either by a local laboratory or by a "central" laboratory which is used by all investigational sites. The results of the laboratory tests are then reported back to the investigational site and/or the sponsoring pharmaceutical company.

7. As the study progresses, case report forms (reviewed by a representative from the pharmaceutical company, as discussed in #5 above) are sent in to the pharmaceutical company. A database is generated, which contains all the data from all the different investigator sites on the protocol entered as different "variables".

8. As data are entered into the database, variables are compared to check consistency of the data within each individual case (for instance, the birthdate of the patient should be in agreement with the reported age of the patient at their clinic visits). Investigator sites are requested to correct any discrepancies. Once all patients have completed the study and all data are entered and believed to be accurate, the database is "locked", meaning that it will no longer be changed. Of course, databases can be "unlocked" again and changes made if major problems are found or if it is otherwise deemed to be necessary.

9. At the end of the study, when all of a site's patients have completed the study, sites are visited again by the pharmaceutical company's representative (described in #5). The purpose of this visit is to close out the site from the study by making sure that all unused drug is collected, the regulatory documentation is in order, and that all remaining case report forms are reviewed and sent in to the sponsoring company.

10. Certain types of data, such as concomitant medications taken during the study and adverse events, may be "coded" to preferred terms or generic drug names, to allow a more consistent analysis of these data across all patients. Once the database is locked and coding is performed, the data are then analyzed across the study, according to the specifications in the study protocol. New analyses can sometimes be added later. Results are presented as figures or in tabular form, and statistical tests are performed.

11. A final study report and/or manuscript is prepared by the sponsoring pharmaceutical company. The purpose of this report, like other scientific reports and manuscripts, is to summarize the main points of the study and to highlight the clinically and scientifically relevant results. These reports usually must conform to federal regulations and guidelines, since the

(continued next page)

Figure 5 (continued):

> reports may be used for a regulatory submission such as a New Drug
> Application.

The process of conducting a clinical trial involves the coordination of many different people with many different roles. For example, the design of the study protocol and case report forms require input from individuals with medical, biostatistical, scientific, regulatory, and marketing expertise. Conduct of a clinical trial often involves the coordination of multiple study monitors across multiple study sites. These people must coordinate with data managers who enter data into the database, and with biostatisticians and scientists and/or medical writers who generate the tables and figures of the data and write the final report. A regulatory affairs specialist is usually on hand to review the report and to be available for regulatory questions, and a project medical officer is available to answer medical questions which may arise during the course of the study. A clinical compliance auditor also audits a select number of M.D. investigator sites to ensure their compliance to the study protocol, regulatory guidelines, and general good clinical practice.

The role of the project manager is to coordinate the entire project to ensure that the appropriate people are talking to one another, and to ensure the entire project is completed according to the company's specifications. At a more advanced level, a product manager coordinates many different projects to ensure that the overall development of the product is going according to the plans of the company at large.

Figure 6
The Clinical Research Process

A Clinical Research Study

Development of Protocol/Case Report Form

M.D.   M.D.   M.D.   M.D.

Data Collected,
In-House Queries Generated,
Data Entered into Computer

Data Analysis and Presentation

Final Report or Manuscript

The New Drug Application

Once a company believes that it has enough supporting information to demonstrate the safety and efficacy of a compound in humans, it then submits a New Drug Application (NDA) to the FDA. Virtually everything that the company has been doing up until this time has been leading toward this application, the approval of which will finally allow the company to begin selling and marketing the product. The purpose of this application is to request federal approval to market and sell the compound in the United States. NDAs are very large, multi-component documents which contain virtually all of the information which is known about a particular compound as it relates to the indication(s) the company is seeking approval to market the product for. These applications are usually focused on one or several different diseases or indications for which the pharmaceutical company wishes to market the product.

New Drug Applications contain meta-analyses of data across studies as well as within individual studies. The purpose of the meta-analyses can be twofold: They can assure consistency across studies, and they can serve as analyses of parameters for which no single study had sufficient numbers of patients to perform a meaningful analysis. For example, a NDA may analyze parameters such as drug-drug interactions between the study drug and other concurrent medications taken by patients for other purposes. It may contain analyses across all studies of the efficacy and/or safety of the drug in a subset of patients who contain a particular type of disease (such as renal or hepatic disease). The types of analyses which are presented in a NDA are determined based on medical and scientific considerations about the compound the indication(s) for which it will be used. NDAs contain data listings of all sorts of pertinent patient information and information about the product's past and proposed manufacturing methods.

Because of all of the information presented in a NDA, they are typically very large and require an abundance of scientific minds and effort! Preparation of a NDA involves clinical scientists, biostatisticians, data managers, regulatory affairs specialists, drug safety specialists, and folks who are involved with the marketing of the product. Sometimes the written portions of the NDA are prepared by a separate writing group, and sometimes they are prepared by the scientists who directed the clinical development program. Other people, such as upper management, are likely to become involved as well. Preparation of a NDA usurps a lot of resources, meaning jobs!

A New Drug Application Can Fill the Back of a Truck!

Phase IV of Clinical Trials

After approval of the New Drug Application by the FDA, Phase IV studies are often conducted. Phase IV studies are aptly named "postmarketing studies" because they are conducted after the drug is already approved by the FDA for sales and marketing within the U.S. Their purpose is to support marketing claims which can increase product sales. For instance, many advertisements state that one drug is superior to another for the treatment of a particular indication. According to FDA regulations, these claims must be substantiated by real data from real clinical trials, which convincingly demonstrate that these claims are justified.

Thus, marketing people are usually involved in the conduct of clinical trials as well as in the assessment of other market parameters (such as competing products or alternative therapies, market size, price the market is willing to pay, and so on).

The Decision to Go Forward

At each step in the pharmaceutical development process, the company needs to decide whether to proceed with the next level of development. A number of different factors can influence this decision. For example, is the product too toxic to be of any potential therapeutic benefit? Does it appear to work? If the drug is not working as intended or is too toxic, the company may choose to engineer the product (for example, by modifying the drug molecule) to try to improve upon it and continue with its development. Alternatively, the company may decide to discontinue development of the product completely.

There are also other factors which influence a company's decision to proceed with the development of a product. For instance, how large is the potential product market? What price will patients be willing to pay for the product? Are there any concerns with the manufacturing process for the product, such as expense of manufacture or important regulatory concerns?

Someone within the company needs to collect and weigh all of these factors and make decisions based on them. This is usually the responsibility of upper management and company executives (with input from lower-level workers). These individuals make their money by making relatively high-stakes decisions about whether to continue developing a product, and if so, how (for example, for what indications, using what drug delivery vehicle, and so forth).

Pharmaceutical companies are in business to make money. Thus, they are constantly evaluating these factors as a means for evaluating the potential profitability of the product. This means that companies are researching the manufacture of the product and evaluating the ways (and costs) of producing the product in very large amounts of sufficient quality, in a cost-effective manner. Additionally, they are evaluating the potential revenues of the product in relation to its development and manufacturing costs. If the company believes that it cannot recover its costs and make a profit, it may choose to stop developing the product.

Regarding the Decision to Go Forward
Important Considerations

• **Product Manufacturing**
Can the product be manufactured in large enough quantities? How much is the cost of manufacture? Can the quality of the product be guaranteed? Since manufacturing of pharmaceutical products is also regulated by the FDA, can the manufacturing process be performed in compliance with FDA regulations and guidelines? These concerns are not trivial, and they can sometimes be the downfall of an otherwise good product.

• **The Potential Product Market**
How big is the market for the potential product? How much will people be willing to pay for the product? What other treatments are available for the disease or indication? How does the potential product measure up to these other alternatives in terms of cost, efficacy, quality of life, and so on? Can the costs of development be recovered by sales of the product?

Product Marketing
    Once a product is marketed, the pharmaceutical company's work is far from over. Companies still conduct clinical trials (Phase IV studies, as described above), to make claims about the product which could increase its desirability and/or patient population. Additionally, product advertising must be designed and implemented. The safety and efficacy of the product must still be tracked, and pharmacoepidemiologic analyses are performed. New formulations, different treatment regimens, and any number of other factors may be studied. The product may also be studied and submitted for approval in foreign countries, such as in Europe or Asia. Foreign countries have different regulations than does the U.S.

Overview of the Pharmaceutical Industry - In Closing
    Now that you've read this section, you may wish to go back to the beginning of the chapter and read the job descriptions in a new and more informed light. You can now relate these job descriptions to the different stages of the pharmaceutical development and marketing process. You can also begin to appreciate the large amount of coordination that the overall process takes, and why there is room for a relatively large number of managerial and administrative positions within this industry.

In Closing - A Few More Hints
    The purpose of the preceding section was to get you started in thinking

about career options in the pharmaceutical industry. You'll find, when you go out into the "real world", that there are many jobs like these. Thus, this chapter will give you a good basis for starting your career exploration in the pharmaceutical industry.

You should, however, be aware that not all jobs will fit exactly into these niches. Chapter 9 discusses the reason for this. Pharmaceutical companies display considerable diversity with regard to size, scientific focus, philosophy, organization, and financial situation.

In some companies, opportunities have also been influenced by the trend toward outsourcing of tasks and entire development programs. As a result, we now see the emergence of "virtual" pharmaceutical companies which consist of only a few individuals who oversee the entire pharmaceutical development and marketing process. Now, pharmaceutical companies can range in size from very small "virtual" companies with only a few employees to very large companies which employ thousands of individuals.

With that, let's take a minute to think about company size. Company size can have a profound influence on the flavor of an organization and the types of opportunities which are available within it. For instance, in the case of a small virtual company, the jobs may only be administrative and managerial in nature, as these individuals will oversee and coordinate the work of their contractors. In larger companies, there will be employees who carry out the work on a day-to-day basis, as well as those who perform strictly administrative and managerial tasks.

Because of the factors mentioned already, individual pharmaceutical companies don't always have a separate person who performs each task of the pharmaceutical development and marketing process. The later chapters of this book (Chapters 9-11) will give you some hints on how to research the situation in specific companies of interest to you. In any event, this chapter, combined with the techniques described in later chapters of this book, can provide you with solid ground upon which you can begin your own exploration of careers in the pharmaceutical industry.

Chapter 3
Biotechnology

The lab is abuzz with its drones,
Producing recombinant clones.
It may seem quite needless
To make something seedless,
But profit will pay back the loans!

There is some old folklore regarding the origins of biotechnology. This folklore maintains that biotechnology was founded based on the following letter and the problem it described.

Dear Miss Manners:

I recently hosted a formal cocktail party at my home in which I served purple grapes. The party went well except for the fact that I was upset by the manner in which people were eating the grapes. What is the proper way to dispose of the seeds?

Some picked seeds from their mouths with their fingers and flicked the seeds into the garbage. Others spit the seeds directly into the garbage. Yet others seemed to be eating the seeds.

Since grapes are a finger food, is it acceptable to bite the grape in half, and pick the seeds out with one's fingers? Alternatively, is it acceptable to put the entire grape into one's mouth and discreetly spit the seeds into a napkin? Is one expected to eat the seeds? I assume that it is not acceptable to spit the seeds directly from one's mouth into a garbage receptacle. Should I have provided some sort of separate receptacle (separate from the regular waste container) to put the seeds into?

- Agape about grapes

While waiting for an answer from Miss Manners, the person who wrote this letter solved the problem. She engineered a seedless variety of purple grapes.

The applications for biotechnology can encompass just about any biological discipline. To prove my point, Figure 7 lists some examples of the types of products and processes for which biotechnology is currently being used.

Biotechnology can be defined as the exploitation of a biological process or product. There are many more examples as well - the uses for biotechnology are virtually unlimited except by our own knowledge and creativity. In any event, for the purposes of this chapter, all companies that develop products or processes based on the exploitation of a biological process or product will be considered biotechnology companies.

Figure 7
Some Types of Biotechnology Products

Products for Human Use
- Gene therapy
- Naturally-occurring or genetically engineered compounds used to treat human disease
- Cosmetic products
- Drug delivery systems (e.g., viral vectors, liposomes)
- Vaccines

Diagnostic/Laboratory Products
- Home diagnostic tests (e.g., pregnancy, HIV)
- Laboratory reagents (e.g., monoclonal antibodies, ELISA kits, plasmids, gene cloning kits)
- Laboratory diagnostic tests (e.g., BRCA-1 test for breast cancer susceptibility, forensic tests such as DNA tests, histocompatibility)

Agricultural Products
- Genetically engineered or selected plants and animals (of agricultural importance)
- Pesticides
- Compounds for use in animals (naturally occurring or genetically engineered-for example, growth factors)
- Vaccines

Industrial Products
- Techniques and methods for large-scale product manufacture (e.g., biological agents, beer and wine, plastics, other organic materials)
- Enzymatic products for various applications (e.g., cleaning)

Environmental Products
- Bioremediation
- Products intended to preserve ecology or to destroy unwanted organisms

In some cases, you may be wondering about the difference between a biotechnology company and a pharmaceutical company. Sometimes the distinction is made based on the method by which a particular compound is manufactured. For example, biotechnology would involve the manufacture of a compound using molecular biological techniques (for example, cloning and expression), whereas a pharmaceutical process would usually involve a chemical synthesis.

Biological agents intended for human use are regulated by a slightly different set of rules and regulations than are pharmaceutical agents (for instance, Product License Applications, or PLAs, are filed in place of New Drug Applications or NDAs). Nevertheless, for most job seekers, you need not worry too much about the distinction. The types of positions are very similar for both types of companies.

So how can you use your science skills to participate in the biotechnology industry?

Career Opportunities within the Biotechnology Industries

As mentioned above, some biotechnology companies are like pharmaceutical companies. These companies may have positions such as those outlined in Figure 8 on the next page.

You are urged to refer to Chapter 2 for a more comprehensive description of the positions shown in Figure 8 and others which relate to the development of compounds intended for human use. In addition, it is important to keep in mind that some biotechnology companies are involved with only limited aspects of the discovery and development of products intended for human or animal use (for example, some biotechnology companies are concerned with development of manufacturing methods or delivery systems for pharmaceutical or biological agents). In these cases, these companies have some, but not all, of the positions described in Figure 8.

Positions Specifically Relating to Biotechnology

The purpose of this section is to provide you with a more in-depth description of career opportunities *specifically as they relate to the biotechnology industry.* Many of these position titles are the same as those presented in Figure 8. However, because there are so many different types of biotechnology companies, each of which has its own mission, products, and set of governmental regulations, the jobs available within a particular biotechnology company may take on a somewhat different focus than those within a pharmaceutical company. Following are some descriptions of positions which may differ from those within the pharmaceutical industry:

•   *Scientific research and development.* Depending on the type of biotechnology company, companies may conduct virtually any type of research. For example, biotechnology companies may focus on environmental research. Other companies may focus on the development of "nutraceuticals" to feed the

Figure 8
Some Career Opportunities in the Biotechnology Industry

- Basic and applied scientific research (including the discovery of new agents or processes, as well as pharmacology, toxicology, manufacturing, and clinical research)

- Regulatory affairs

- Biostatistics

- Product manufacturing

- Quality control and quality assurance

- Database management

- Administration and management of the organization, scientific directions, and its personnel

- Product marketing

- Product sales

- Product licensing and technology transfer

- Epidemiology

growing population of the world. Other companies may focus on the development of assay systems or new laboratory reagents.

• *Regulatory affairs.* The activities of many biotechnology companies are regulated by government agencies other than (or in addition to) the Food and Drug Administration (for example, the Environmental Protection Agency, the United States Department of Agriculture, and the Occupational Safety and Health Administration, which regulates worker safety). Regulatory affairs professionals in these industries need to be familiar with different sets of regulations than those which pertain to the pharmaceutical industry.

• *Epidemiology.* Biotechnology companies track the cause and incidence of diseases for many different reasons. In addition to tracking human disease, some companies may track animal and plant diseases for such purposes as the development of products designed to protect crops or livestock. Thus, the focus of epidemiology can differ from company to company.

• *Data management and computer programming.* There are many reasons for biotechnology companies to keep databases. Some of these are similar to those of the pharmaceutical industry, and some are not. For instance, biotechnology companies may analyze data from environmental, toxicological, pharmacological, preclinical, clinical, epidemiological, and basic scientific types of studies. The type of knowledge that is required for successful creation of a database, its management, and analysis depends very much on the type of biotechnology company and what they are planning on using the information for.

• *Product marketing.* Like the pharmaceutical industry, biotechnology companies often conduct market research. However, due to the different nature of their products and markets, a job in marketing for a biotechnology company may be quite different from that of a pharmaceutical company. For example, a company which sells laboratory reagents will not conduct the same types of marketing research as a company which sells a pharmaceutical agent.

• *Product sales.* Biotechnology companies, like pharmaceutical companies, often have salespeople who will try to sell the products to different customers. However, the sales approaches used may differ significantly from company to company, depending on the product and its market. Sometimes biotechnology companies have their salespeople visit people in their markets (for instance, laboratories, hospitals, veterinary facilities, environmental or agricultural firms, and so on), whereas in other companies, sales may be conducted over the telephone.

• *Quality control and quality assurance.* Quality control and quality assurance measures can depend on the type of biotechnology company and the products and/or processes it sells. Some audits and inspections may be performed to fulfill the regulations of the federal or local government (again, depending on the type of industry), whereas others may be performed simply to assure product quality to customers.

• *Technical writing.* In addition to the technical writing which is performed by pharmaceutical companies and some biotechnology companies for submission to regulatory agencies, technical writing may be performed for a variety of other purposes. For instance, biotechnology companies may prepare technical brochures or manuals which describe their products or their use.

• *Product manufacturing.* Positions in product manufacturing can depend very much on the type of biotechnology industry and the type of product that is being manufactured. Some products (such as biological agents for human use) are very tightly regulated by one or more government agencies, whereas other products (such as cell cultures) are less so. These differences, as well as the methods of product manufacture and the ultimate use of and packaging of the product, can have important implications for the nature of manufacturing jobs.

• *Customer Service and Technical support.* Some types of biotechnology companies make products (for example, diagnostic tests or laboratory reagents) which require some technical expertise in order to use them. Companies which develop products of this nature often have a person who is available to answer questions from customers regarding the product.

In Closing

When evaluating the types of positions available within the biotechnology industry, you will probably find it helpful to bear several thoughts in mind. First, consider the type of biotechnology industry in which you are interested. What types of products does it make and sell? Second, consider the size and scope of the company. What work does the company do, and does it do all of its own work (as opposed to outsourcing some of the work)? Biotechnology companies, like pharmaceutical companies, can range in size from large, self-contained companies to "virtual companies" which consist of only a few individuals. They can either do most of their work in-house or outsource it. Finally, find out as much as you can about the company's organization in relation to the tasks which need to be accomplished. This topic is discussed further in Chapter 9. Then think about how your skills might fit in to this overall organization to make a contribution to it.

If you are interested in a career in biotechnology, you are also urged to refer to Chapter 10 on how to keep current with opportunities within your field. Biotechnology is constantly changing as it is applied to different problems and as new technologies and discoveries are introduced into it.

Chapter 4
The U.S. Federal Government

The federal budget is tight,
Our debt has gone way out of sight.
You might be less nervous
If you were in service,
Ensuring the job was done right.

What do biologists in the federal government do?

Don't wait too long for the punchline. This is not a joke. I know that some people believe that "government worker" is an oxymoron, but in reality, there is ample evidence to suggest that government workers actually do work.

Just think about it in relation to your taxes. If you owed money to the IRS, would you actually file a return and ante up, even if you thought you wouldn't get caught for not doing so? Maybe YOU would. However, there are many people who wouldn't.

However, people DO pay. They DO believe that they might get caught. Thus, they must believe that government workers work.

In fact, many federal government workers work very hard. In the biological sciences, they administer grants, conduct research, and make and enforce laws, policies, and regulations. They advise the president. They provide information, resources, and other services to the scientific community and to the general public. They also do other things, some of which are described in this chapter.

Positions within the U.S. Federal Government

The following positions are examples of the types of work you can find within the U.S. federal government. It is important to remember, though, that not all government organizations carry out the same type(s) of work; thus, not all

Figure 9
Some Career Areas in the U.S. Federal Government

- Scientific Researchers

- Other Laboratory Workers

- Hospital Workers

- Technology Transfer

- Auditors and Inspectors

- Reviewers of Technical
  Materials / Applications

- Policy

- Economic Impact /
  Budget Analysis

- Database Managers

- Biostatisticians

- Writers and Communicators

- Administrators and Managers

- Technical Lawyers

organizations have the same types of positions, or the same position titles. Please refer to the United States Government Manual (see Appendix C) and/or to Chapter 10 to learn more about the different organizations of the federal government and the types of work that they do. Furthermore, some examples of government organizations which may be of particular interest to scientists are covered later in this chapter.

There are several different employment arrangements which are available within the federal government. These include permanent civilian employment, civilian contract employment, and enlisted workers within the U.S. military. Permanent civilian employment is more-or-less similar to the "permanent" employer-employee relationships seen in the private sector. Contract employment is described in more detail in Chapter 7. Enlisted workers are actually in active military service, either as members of the "Commissioned Corps" (which exist to provide medical care to the U.S. military in time of war), or with the Army, Air Force, or Navy. Enlisted workers can be called to service at any time in the event of war.

Many government positions are available for any or all of the different types of employment arrangements described above.

Scientific Researchers
The federal government offers many positions in scientific research, both at the bench and in other venues, and in virtually every scientific area and discipline. For example, the research can focus on human, animal, or environmental health. It can be applied or more basic in nature. It can involve

research with humans (for example, to test the effect of certain environmental conditions, such as those which may be encountered in space travel), or it can involve the gathering of epidemiological, toxicological, and/or statistical data. It can involve the development of new forensic techniques. The types of research conducted by the federal government may be used to help make governmental policies, to decide on areas of interest for federal grant funding, to write regulations, and to make decisions regarding the approval of industrial products.

Because governmental research may be used for federal decisionmaking or for other federal purposes, research scientists within the government sometimes have additional responsibilities besides their research. For example, they may review drug applications at the Food and Drug Administration (FDA), or they may hold academic appointments at the Uniformed Services University of the Health Sciences (USUHS). Nevertheless, there are some scientists within the federal government who can do research on a more-or-less full-time basis.

Other Laboratory Workers

Besides conducting basic scientific research, there are other opportunities for laboratory workers within the federal government. For example, the federal government has forensics laboratories which conduct tests for criminal investigations, and clinical laboratories for their hospitals and other medical testing. Additionally, the federal government conducts numerous types of inspections and audits which may warrant testing of samples in the laboratory. For more information on the types of audits and inspections that the government conducts, please see the section on audits and inspections below.

Hospitals

The federal government supports hospitals, such as Veterans Administration hospitals and other hospitals for those who are wounded and/or permanently disabled during war. These hospitals need doctors, nurses, laboratory technicians, and researchers to conduct the work and perform the missions of these hospitals.

Technology Transfer

Like other institutions that conduct research, some of the agencies within the federal government may want to in-license patented inventions for their own purposes. Additionally, the federal government may out-license inventions that have resulted from its own research. The process of buying or selling the rights to use a particular invention for a particular purpose is known as "technology transfer". Technology transfer is also described in other chapters of this book.

Technology transfer involves a knowledge of the scientific purposes for which the product or invention is being licensed, as well as a knowledge of how to execute the contractual agreements which allow the actual technology transfer. Additionally, individuals who work in technology transfer can assist government workers with finding outside organizations and/or individuals who are interested in entering into potentially useful licensing agreements.

Auditors and Inspectors
Auditors and inspectors travel to businesses and research sites to ensure compliance with federal regulations. For example, these individuals may inspect businesses for compliance with regulations such as paperwork requirements (for example, FDA), worker safety (for example, OSHA), or environmental emissions (for example, EPA). In each of these cases, the job title of the auditor or inspector depends on the governmental organization he or she works for, as well as the nature of the job that he or she inspects or audits. In some cases, these auditors or inspectors also perform laboratory procedures to test compliance, whereas in other cases, the labwork is either not required or is done by a separate individual. Sometimes, reviewers of technical applications may direct and/or participate in these audits and inspections, as is often the case at FDA.

Reviewers of Technical Materials/Technical Applications
The federal government reviews several different types of technical applications such as applications for new drugs, biological agents, and medical devices (which go to the FDA), and patent applications (which go to the U.S.Patent and Trademark Office). These reviews can be very multidisciplinary (for instance, at the FDA, where applications can undergo medical, statistical, pharmacology, environmental, chemical, microbiological, and other reviews). Thus, the federal government hires scientists with expertise in each of these areas to review these applications. These technical reviewers may be involved in other activities as well, such as research, or they may review these technical applications full-time.

Policy
One of the ways in which the federal government protects the public is by making policies regarding various issues. Many of these policies are related to biological concerns such as ensuring a safe food supply, safe and efficacious drugs, protection of the environment for human and animal health, both for the present and the future. In order to assess the potential and actual effectiveness of these policies, the federal government needs individuals who are able to understand the scientific principles pertinent to the policies. Armed with this knowledge, these individuals can then identify important scientific policy issues, develop policies, and analyze their effectiveness.

Economic Impact and Budget Analysis
In making decisions about its various policies such as research grant funding, the prevention or control of disease, and protection of the environment, the federal government takes cost, among other factors, into consideration. For instance, government workers may analyze the projected costs of a disease, such as the incidence of the disease, the cost of health care due to the disease (both short- and long-term), and whether the disease causes workers to quit working

(thereby losing wages). This type of information can help the federal government to determine the economic impact of a particular type of disease. This information, along with scientific judgment (such as the likelihood that research will lead to a cure for the disease, and its effect on quality of life), can be used to make decisions about whether to fund additional research on a treatment for the disease, to attempt to control it by increasing funding in some other way (for instance, by preventing the disease in the first place), or to limit spending on this disease in favor of a more pressing health concern.

In addition to these types of analysts (and the policy analysts mentioned above), the federal government has analysts who examine other parameters such as the budgets for particular government programs. These analysts examine how much the implementation of a government program costs, or is likely to cost, in all of its component parts. These individuals try to tell the government whether a particular program is on-budget, and if not, why they think it is not. These individuals can suggest ways to reduce spending on government programs, advise the government about budget issues, as well as to suggest ways to cut program budgets.

Database Managers

The federal government, like many other organizations, maintains all sorts of scientific databases. For example, the government may have databases which contain epidemiologic information, gene sequence information, results of toxicology, environmental, or other scientific studies, and so on. In some cases, these databases are made available to the public. In other cases, they are used for analysis by a government agency or some other organization.

In any event, the federal government needs individuals who can understand the scientific purpose of the database in sufficient detail to generate a usable and meaningful database which can be grouped, understood, and/or analyzed. Database managers must therefore have a sufficient understanding of science to communicate with the various scientists (for example, epidemiologists, toxicologists, biostatisticians and/or others) for whom the databases are intended.

Biostatisticians/Biostatistical Programmers

Like many other organizations, the federal government often needs statistical input in the design of studies (either its own, or those of others), the type of information that needs to be collected for statistical testing, sample size, and the types of statistical tests with which the information can be analyzed. As in other organizations, biostatisticians in the federal government work in conjunction with other scientists and database managers to design and analyze study endpoints. Biostatisticians also participate in the review of technical applications (for example, to the FDA) with regard to their biostatistical content and implications.

Writers and Communicators

One of the important functions of the federal government is to communicate with other government organizations and with the public regarding its policies, regulations, research, and other work. For instance, the government may wish to prepare educational materials for the public, or for other government organizations about certain diseases, drugs, environmental concerns, or whatever. Government educational programs may involve the preparation of informational pamphlets, or they may involve educational talks to the public. They may also involve the news media. In some cases, the writing and communicating is done in conjunction with another job (such as research), whereas in other cases, it can be a full-time job.

Administrators and Managers

The federal government, just like any other organization, needs individuals to administer and manage its programs and workers. These individuals oversee the overall direction of the programs and budgets for which they are in charge, and they coordinate the tasks and facilitate the communications which are required for completion or implementation of a particular program.

For instance, there are individuals who manage and administer federal granting programs. There are individuals who administer programs for the prevention or control of disease. There are individuals who administer the overall functioning of a particular government agency. Effective management of these programs often requires the administrator to liaise with government workers, outside reviewers, and/or individuals or organizations for whom the programs are intended.

Obviously, the nature of these administrative and managerial positions depends on the nature of the government agency. Every organization needs workers who perform these types of roles. A knowledge of the missions and programs of the agencies in which you are interested will help you to understand the specific positions which may be available to you. Nonetheless, there are numerous positions of this type in virtually all of the different organizations of the federal government.

Technical Lawyers

As you well know, the federal government is responsible for lawmaking, some of which relates to scientific or technical issues. For example, the government writes regulations and makes policies regarding medical, environmental, and other industrial issues. It also makes decisions regarding patents of scientific processes and products.

The federal government therefore has lawyers with technical as well as legal expertise. These lawyers can suggest policies and offer advice and opinions about both the legal and technical implications of a particular course of action, law, or regulation.

Subdivisions within the U.S. Federal Government Which Are Likely to Be of Interest to Biological Scientists

The Federal government is a very large and complex organization. It has many different departments, offices, agencies, and other subdivisions and organizations. Each one of these subdivisions serves a particular purpose. For instance, different subdivisions are concerned with agriculture, pharmaceuticals, and the environment. Some subdivisions have regulatory authority, whereas others exist mainly for the purpose of advising lawmakers or for conducting research.

It is important to understand that the U.S. government is a dynamic organization which changes along with shifts in public policy, public opinion, and law. New subdivisions can be formed, and others can be disbanded. Positions are created or deleted as they either become needed or obsolete. Thus, an understanding of the types of positions which are available in the federal government requires that you be aware, at least to some extent, of current events in the areas of your interest.

Luckily for us, there is a publication which comes out every year which describes the structure and function of the U.S. government in that year. It is the United States Government Manual. This publication contains descriptions of the different subdivisions of government and the missions of each of these subdivisions. It is available at many libraries and can be obtained by contacting the address or phone number provided in Appendix C in the back of this book.

Although the U.S. government is always in a state of flux, I will nevertheless attempt to describe just some of the organizations of government which may be of interest to you. Many of these government organizations have existed for a while and are likely to remain at least somewhat stable for the foreseeable future. I have included these descriptions for the sake of example, so you can gain insight into the types of work that the government does. However,

Figure 10
Some Government Organizations
with Biological and/or Biomedical Focus

• Department of Health and Human Services *National Institutes of Health (NIH) *Centers for Disease Control (CDC) *Food and Drug Administration *Other	• National Aeronautics and Space Administration • Department of Energy • Office of Science and Technology Policy
• Department of Agriculture *Agricultural Marketing Service *Animal and Plant Health Inspection Service *Food Safety and Inspection Service *Agricultural Research Service *Economic Research Service *Other	• Department of Defense *Army *Navy *Air Force • Department of Commerce *Patent and Trademark Office *Technology Administration *National Oceanic and Atmospheric Administration
• Department of the Interior *Fish and Wildlife and Parks Division *National Park Service *Other	
• Environmental Protection Agency	• Department of Labor *Occupational Safety and Health Administration
• National Science Foundation	• Federal Bureau of Investigation

this information is only as current as the writing of this book. Furthermore, these descriptions are merely paraphrases of the official mission statements of these organizations and may, as such, be somewhat misrepresentative and/or inaccurate. Thus, you are urged to refer to the United States Government Manual for the more complete and accurate information, as well as for information on where to apply for employment for each of these different organizations.

Department of Health and Human Services

The Department of Health and Human Services has a large number of organizations which are likely to be of interest to biological scientists. Many of these organizations are located within a subdivision of the Department of Health and Human Services which is known as the Public Health Service. Following are a few of these organizations.

The **National Institutes of Health (NIH)** conducts reseach on topics related to human health and promotes the dissemination of knowledge related to these topics. As you probably know, NIH also administers grant programs which fund extramural research (for instance, at universities) and training programs for new scientists (many scientists have competed for these grants). As its name suggests, NIH is actually a composite of separate institutes, each of which focuses on a particular biomedical topic area.

The **Centers for Disease Control (CDC)** conducts research and compiles information and statistics on the epidemiology of disease, at both the national and international level. Its ultimate purpose is to promote the control and prevention of diseases. The CDC administers public programs for the dissemination of information and the prevention of diseases, whether infectious or caused by other factors such as diet or exercise. Like NIH, the CDC consists of multiple organizations, each of which focuses on a different function.

The **Food and Drug Administration (FDA)** administers programs to monitor the safety of foods, drugs, medical devices, and cosmetics. It does this by creating regulations and guidelines, reviewing applications, performing research, and conducting inspections. The FDA also disseminates information about the products that it regulates. Like NIH and CDC, the FDA consists of multiple subdivisions which are involved in the different missions of FDA.

There are other organizations within the Department of Health and Human Services which may also be of interest to biological scientists. For instance, the **Agency for Toxic Substances and Disease Registry** monitors and compiles information regarding human exposure to toxic substances. This agency then disseminates this information and makes policies regarding exposure to these toxic substances. The **Agency for Health Care Policy and Research** conducts research and disseminates scientific information about the quality and cost-effectiveness of health care. The **Health Resources and Services Administration** is concerned with the accessibility of health care, as well as its quality and cost.

United States Department of Agriculture

Like the Department of Health and Human Services, the United States Department of Agriculture (USDA) is a highly complex organization which contains many organizations which are likely to be of interest to biologists. These include the **Agricultural Marketing Service**, the **Animal and Plant Health Inspection Service**, the **Food Safety and Inspection Service**, the **Agricultural Research Service**, the **Economic Research Service**, the **Cooperative State Research, Education, and Extension Service**, the **Food Nutrition, and Consumer Services**, the **Farm and Foreign Agricultural Services**, the **Forest Service**, the **Natural Resources Conservation Service**, and numerous other organizations. Please see the United States Government Manual for more information on the missions of each of these organizations of USDA.

Some of the activities of USDA which may be of particular interest to biologists include the USDA's numerous research programs, as well as the administration of programs and regulations to monitor the quality of food and veterinary products. For instance, the USDA has programs to test and inspect food to ensure its quality and freedom from harmful contaminants. It also administers programs and regulations regarding animal and plant health, the genetic engineering of plants and animals used to produce food products, and the use of veterinary biological products in animals. The USDA also conducts research to develop new agricultural technology and works with other agencies and institutions to provide resources and access to scientific knowledge. There are numerous other activities of the USDA as well; you are urged to refer to the United States Government Manual for a better understanding of the USDA and the different services it performs. The United States Government Manual also describes the different organizations within USDA and each of their missions.

Environmental Protection Agency

The Environmental Protection Agency (EPA) conducts research on environmental issues, develops programs, and creates regulations and guidelines for environmental protection. The EPA also has programs to monitor and inspect the condition of the environment, as well as to monitor environmental emissions. The EPA has multiple subdivisions, each of which is concerned with a different aspect of environmental protection.

National Science Foundation

The National Science Foundation (NSF) administers scientific programs and recommends policies and actions to promote scientific research and education. Additionally, NSF administers grant funding to extramural institutions for basic scientific research and education.

Department of the Interior

The Department of the Interior exists, at least in part, to protect and preserve our wildlife, resources, fish, and land. Within it is the **Fish and Wildlife and Parks** Division, which includes the **National Park Service**, the **National Biological Service**, and the **U.S. Fish and Wildlife Service**. The **Water and Science** Division, which develops and implements water, mineral and science policies is also likely to be of particular interest to biologists. The **Bureau of Land Management** and the **Bureau of Reclamation** may also be of interest to biologists.

Department of Defense

The Department of Defense conducts research which is relevant to our national defense. Research is conducted by the **Army**, **Navy**, and **Air Force**. For example, research is conducted on diseases, such as infectious diarrhea, which are prevalent among soldiers during wartime. Research is also conducted to prepare for the event of germ and/or chemical warfare, and to understand the effect of certain conditions which may occur during war (such as fast airplane accelerations) on man. This type of research can aid in the design of military equipment.

The **Uniformed Services University of the Health Sciences** (USUHS) is a government-run university whose purpose is to educate medical officers for military service. USUHS also conducts scientific research.

Department of Energy

The Department of Energy (DOE) administers programs for the use and safe disposal of energy resources (for example, radioactivity). It also has programs for national security, environmental management, and science and technology.

With regard to the biosciences, the DOE administers and funds many different types of programs - for instance, workers may study the effects of energy (such as radiation) on biosystems. Biological systems may also be used to fulfill the DOE's missions of finding safe and economical sources of energy (for example, photosynthesis). The department is concerned with the effect of these energy sources (and the waste generated by their utilization) on the environment, the ecosystem, and on the human population. Thus, the DOE administers programs involving each of these aspects of energy utilization and disposal.

National Aeronautics and Space Administration

The National Aeronautics and Space Administration (NASA) conducts research regarding the effect of travel in and out of the Earth's atmosphere on people and other biological systems. It also conducts experiments in space, which are designed to answer questions such as the effect of conditions which exist in space, such as weightlessness, on biological systems.

Office of Science and Technology Policy

The Office of Science and Technology Policy advises the president with regard to scientific policies, plans, and programs of the federal government. Thus, this office analyzes issues of scientific concern and influences national policy on the scientific research and development programs of the federal government.

Department of Commerce

The Department of Commerce makes policies and offers services regarding science as it relates to business. For instance, the **Technology Administration** plays a role in technology policymaking and serves as a resource for scientific and other business-related information. It also provides access to many different government documents regarding technology, provides services in the area of technology transfer, and helps industry with the development of standards, tests, and technology. The Technology Administration has several subdivisions which may be of interest to biological scientists; you are urged to see the United States Government Manual for more information about them.

The Department of Commerce also has several other organizations which may be of interest to biological scientists. For instance, The **Patent and Trademark Office** reviews patent applications for biological and pharmaceutical products, manufacturing processes, and other inventions, and the **National Oceanic and Atmospheric Administration** conducts research and provides services relevant to the preservation of marine life.

Department of Labor

The Department of Labor is interested, among other things, in ensuring the safety of workers in the U.S. This is an important function of the **Occupational Safety and Health Administration (OSHA)**. OSHA sets standards for worker safety and inspects workplaces regarding exposure of workers to potentially harmful conditions such as dangerous machinery and chemical and biological agents.

Federal Bureau of Investigation

The Federal Bureau of Investigation (FBI) investigates crimes. The examination of scientific forensic evidence is part of the FBI's activities. The FBI runs forensic laboratories and also has agents who are trained in science and who can identify and handle scientific evidence properly.

These government organizations are just some examples of potential employment opportunities for scientists. Depending on what you want to do, you will find opportunities in other areas of government as well. Please refer to the United States Government Manual for a more complete description of the U.S. federal government.

Chapter 5
State and Local Governments

You want to determine the fate
Of neighborhoods, city and state.
You hope to save trees,
Or wipe out disease-
Or maybe you like to debate!

When it came to learning about scientific jobs within state and local governments, Pam learned the hard way. Here's how.

It was a hot and sticky Saturday morning, the seventh day in a row of intolerably hot and humid weather. Pam had slept fitfully, if at all. Now soaked with sweat, she wiped her eyes and tried to think about the day ahead of her.

This heat wave was unusual for this northern town. The mercury in Pam's apartment had already topped 95 degrees and was climbing. Because the apartment had no air conditioning, there was nothing she could do about it but turn on her fan and keep sweating.

Desperate to escape into the world of air conditioning and fantasy, Pam had gone to the movies for the last three nights in a row. Today she contemplated spending the day in the waters of cool Lake Ontario. She staggered outside to fetch her morning paper. In doing so, she decided that yes, definitely she would spend the day at the beach, taking respite in the water and shade under one of the umbrellas.

She picked up her paper and scanned the headlines. Then she saw a headline that would have an impact both on this particular day and on her life:

"BEACH CLOSED DUE TO HIGH COLIFORM COUNT"

Pam was irate. How dare "they" close the beach on one of the hottest days of the year? She stormed back into her apartment and back outside again, unable to believe what she had just read. Obviously the person who had made that decision had air conditioning.

She read on: "Due to some unusual weather patterns over the last several days, the number of coliform bacteria in the waters of Lake Ontario's local beaches has increased to dangerous levels. Beaches will be closed to swimmers until further notice.

"Coliform bacteria live within the intestinal tract of animals and humans. They are indicators of fecal contamination in water. When coliform counts soar to certain levels, beaches are closed because the water is considered unsafe for swimming.

"All beaches will remain closed until further notice."

Pam read the article again. "Who was making these decisions, anyway?", she thought. Boy, would she like to talk to them! At the very least, she'd like to find out their names so that she could be sure not to vote for them again!

Then she had an idea. Pam decided to spend the day at the local library. After all, the library was air-conditioned! It was also a place where she could research the types of jobs that allowed a person to close beaches on the hottest days of the year.

After taking a cool shower, Pam headed for the library. She soon struck gold. She learned that her state and local governments actually employ a lot of scientifically-minded individuals to make these types of policies and decisions. In fact, they can do much more.

Positions within State and Local Government Offices

There are lots of jobs for biological scientists within both state and local governments. The following section gives you some examples of what the nature of these jobs might be.

When reading this chapter, you'll want to keep in mind that the specific types of jobs available in any one region, state, or local government will depend on many factors. Obviously, there are 50 state governments, and many more local governments. Each of these regions and governments has its own geographic and scientific concerns, public and political climate, empowerment, and tax base.

Nevertheless, what are some of the scientific issues which face state and local governments? There are programs in human health and safety issues, bioethics, environmental preservation, and agriculture, just to name a few. All of these types of programs need scientists who are familiar with the issues and who can make decisions regarding them.

This chapter is designed to introduce you, in general terms, to some of the types of positions which you may be able to find in the state and local governments. However, since there is so much variation from one state or local government to another, you are urged to refer to Chapter 10, once you have read this chapter, for some hints on how to become familiar with some of the issues in the community of interest to you. This knowledge will then help you to identify specific opportunities within a particular state or local government.

Figure 11
Some Careers in State and Local Governments

- Scientific Research

- Other Laboratory and Technical Positions

- Auditors, Inspectors, and Other Law Enforcement Officials

- Educators and Communicators

- Policy

- Economic Impact and Budget Analysis

- Computer Database Specialists and Computer Programmers

- Biostatisticians

- Administrators and Managers

Scientific Research

Many state and local governments conduct basic scientific research, both at and away from the laboratory bench. For example, researchers and epidemiologists may track the incidence and etiology of local diseases. They may study the bacterial or viral strains, environmental substances, and/or other factors involved in a disease outbreak. They may conduct research to identify new biologically-based forensic techniques to be used in the laboratory or at a crime scene. They may study the effect of industrial emissions on the local environment.

The research programs conducted by state and local scientific researchers can be either basic or applied in nature. They may be supported by federal money, instead of or in addition to state and/or local money. In fact, many state and local government researchers apply for federal grant money in the same way as an academic researcher does (see Chapter 6).

Other Laboratory Positions

State and local governments often conduct laboratory tests to monitor environmental conditions, to diagnose and/or track diseases, to identify perpetrators or victims of crimes, and much more. These laboratory tests may or may not be a part of a research program. For example, laboratory workers may routinely test the coliform count in beachwater to decide whether the water is safe for swimming. Workers may routinely test all clinical specimens of a particular type for a particular microorganism or toxin.

Auditors, Inspectors, and Other Law Enforcement Officials

State and local governments may work either alone, together, and/or in conjunction with the federal government, to ensure that local businesses, laboratories, and other institutions are compliant with federal and/or local regulations. For example, auditors or inspectors may monitor an institution's facilities, operating procedures, records, environmental emissions, products, or other policies, to ensure that they meet the rules or standards of quality which apply to them. Some of these auditors and inspectors may also conduct laboratory tests on samples they collect. At other times, these samples may be sent in to a separate laboratory for testing.

It is noteworthy that the individuals who are involved in these types of audits and inspections are actually ensuring compliance with the law. Violations of these laws may be a criminal offense. In keeping with this concept, some of the individuals who conduct these types of audits and inspections are actually law-enforcement officials. That is, these individuals have the power to make arrests when a person and/or company is found to be in violation of the law.

Education and Communication

Most state and local governments have educational programs for the public. These programs can be designed with the intent of educating students, preventing disease, protecting the environment, for public interest, or for a variety of other purposes.

State colleges and universities are one example of educational programs administered by the state. The workers within these universities are actually employees of their state governments. Many state and local governments also support museums, nature preserves, parks, arboretums, and other exhibits of science and nature. These museums need interpreters and educators to design and explain displays, to prepare pamphlets, to give talks and lead nature walks, and so on. State and local governments may also design educational materials such as pamphlets, displays, and talks for other purposes, such as for the prevention or control of disease or protection of the environment. Thus, educators and communicators may be writers, as well as speakers, interpreters, and/or public relations people.

Policy

Like the federal government, state and local governments may have individuals who make and break policies regarding biological issues. These issues can involve human health, the environment, agriculture, or any number of other regional concerns. People who work in policy are responsible for identifying the important policy issues of a particular region, as well as designing and implementing policies, and devising methods for analyzing the effectiveness of these policies. Since the implementation of effective policies requires some understanding of the scientific issues revolving around a particular problem or concern, these types of positions often exist to educate and advise people (such as lawmakers) about the scientific implications of a particular policy. These types of positions may require considerable training within the biological sciences.

Economic Impact and Budget Analysis

In making decisions about its various policies, state and local governments (perhaps in conjunction with the federal government) consider the cost of programs in regard to their overall economic and/or social benefit. Thus, government workers may analyze the projected costs for the implementation of any particular program in relation to the projected costs of doing something else, or in comparison to the social and/or economic costs of doing nothing at all. This topic is also discussed in Chapter 4.

Once programs are implemented, government workers analyze the program budget, to determine whether the program is staying on track financially. Workers may also analyze the effectiveness of the program in terms of the cost savings to the individuals, institutions, or community for whom the program was intended.

Although these positions can be mainly financial in nature, scientific judgment sometimes does come into play. For example, it is not always clear as to whether a cost or saving can be attributed to a particular program, or whether it should be included in a particular type of analysis. Scientific judgment may also be required for deciding on the size of the budget which is necessary to implement a successful program.

Computer Database Specialists and Computer Programmers

State and local governments, just like the federal government and many other organizations, maintains all sorts of scientific databases. For example, the government may have databases which contain epidemiologic information, results of toxicology, environmental, or other scientific studies, and so on. Some of these databases may be intended for the public, whereas others may be intended for scientists, lawmakers, policy or budget analysts, or for any number of different individuals.

In any event, state and local governments need individuals who can understand the scientific purpose of the database in sufficient detail to generate a usable and meaningful database which can be grouped, understood, and/or analyzed. Database managers must therefore have a sufficient understanding of science to communicate with the various scientists (for example, epidemiologists, toxicologists, biostatisticians and/or others) for whom the databases are intended. These institutions also need individuals who are sufficiently skilled in computer programming as well as in science, to actually perform the database analyses.

Biostatisticians

Like many other organizations, state and local governments may need statistical input regarding their studies and the analysis of data. Biostatisticians can help with decisions about what types of data need to be collected, as well as how much. They can provide the expertise necessary to do statistical comparisons of data.

Administrators and Managers

Just like every other organization, state and local governments need individuals to administer and manage its programs and workers. These individuals oversee the overall direction of the programs, institutions, and/or budgets for which they are in charge. They coordinate the tasks and facilitate the communications which are required for the completion or implementation of a particular program. Sometimes they liaise with federal officials as well.

There are many different types of administrators and managers. Chapter 2 explains some of the different types of administrative and managerial positions which exist within the pharmaceutical industry. Although the focus of state and local governments is different than that of industry, the concepts remain the same. Thus, you are urged to refer to that chapter for some examples of the different types of administrative or managerial positions which exist.

For instance, there are individuals within state and local governments who manage personnel and contractors. There are individuals who manage and implement programs for the control of disease or preservation of the environment. There are individuals who manage research grants and who decide on the distribution of funding among the government's various programs. There are also individuals who oversee the various projects within a particular program.

Obviously, the actual job title of these administrators and managers depends on what types of programs they are administering and/or managing. Nevertheless, for those people who are interested in this type of work, there are many opportunities for scientists.

Chapter 6
Academic Institutions

You get in the classroom to teach,
The students you try to beseech.
You fight hard for grants,
But when there's a chance,
You summers are spent at the beach!

Susie began preparing for an academic career at a very early age. The scenario usually went something like this.

The doorbell rang.

"Who's there?", the resident asked.

"It's Susie from next door. I'm selling candy bars to raise money for a band trip this coming fall. Would you like to buy some?"

"How much are they?", the resident asked.

"A dollar a bar", Susie replied.

"I would like to buy one, but unfortunately, I don't have much change in the house right now", the resident said. "Can you come back tomorrow?"

"Sure", Susie said, "thanks."

After her third visit to the house, the resident finally bought a candy bar.

Now let's see how that relates to a career in academic science. Let's flash ahead 25 years.

The Federal Express package arrived. The secretary opened it.

"Dear National Institutes of Health", the cover letter started, "Enclosed please find my grant application to support my research on the induction of the melanin gene in humans."

The secretary put the application in the appropriate pile.

The study section met to review the application.

"How much money does the principal investigator want?", a study section member asked.

"A million dollars over a 5-year period," another member responded.

"Well, it's for a good cause. The science is good, too. However, we don't have the money to fund all of these grant applications right now. This one'll probably have to reapply again in the fall."

After about the third try, Susie finally obtained her grant funding.

Little did Susie know how well her door-to-door candy sales of childhood would teach her the patience and persistence she would need in order to obtain grant funding for her research.

Positions within Academia

Actually, there are many different employment arrangements which exist in academia. The exact nature of the positions available depends very much on the type of academic institution for which you want to work.

Let's start by reminding you of the different types of academic employers for whom you can work:

- Grade or Secondary Schools (Public or Private)
- Two-Year Colleges, Technical, and Trade Schools
- Four- or Six-Year Colleges
- Universities

Of these different types of institutions, universities alone have full-scale research programs. Some four- and six-year colleges may also have research programs, but they tend to be more limited in scope. Nevertheless, there is more than one type of employment arrangement for scientists within each of these institutions.

The following are just some of the types of work you can find in academia. Obviously, not all of these positions exist within every academic institution.

Figure 12
Some Academic Positions

• Teaching	• Program Administration and Management
• Research	*Grants
• Teaching and Research	*Technology Transfer
• Academic and Teaching Hospital Positions	*Library and Information Science
	*Public Education
	*Career Counseling
• Laboratory Technicians	*Health Physics / Chemical Safety (e.g., safe use of radioactivity and lab chemicals)
• Clinical Study Coordinators	
	*Ethics Programs (e.g. animal and human rights)

Tenure-Track Faculty Positions

Everyone is familiar with the existence of tenure-track faculty positions. These are the positions that so many people in academia strive for. Tenure-track faculty positions usually involve either teaching alone (in teaching institutions), or a combination of teaching and research (in universities and some colleges).

In keeping with the concept of academic freedom that tenure is supposed to embody, the types and styles of teaching and research which occur within academia can be very diverse. Some tenure-track faculty teach mainly in the classroom and conduct basic scientific research in their laboratory, on topics of their choosing (as long as they can obtain funding for it). Other tenure-track faculty may teach in a more applied setting, such as in a university-based hospital or clinical laboratory, in addition to their research (again, usually on topics of their choosing, for which they either obtain their own funding or work in conjunction with other scientists).

Tenure-track faculty salaries are usually paid by the institution at which they are employed. However, many faculty in research institutions are expected to obtain sufficient funding to pay for at least a portion of their salaries.

Non-Tenure-Track Teaching Faculty Positions

Non-tenure-track faculty may involve teaching, research, or both. Teaching positions are usually on an "instructor"-level basis. In these cases, the instructor's salary is usually paid by the institution. Instructors in colleges and universities may also decide to perform research, either with their own grant funding or in someone else's laboratory. Usually, though, an instructor is not required by the institution to perform research.

Non-Tenure-Track Research Faculty Positions

Non-tenure track research faculty can either direct their own laboratories and conduct research on topics of their own choosing (with their own funding), or they can work in someone else's laboratory (in which case, they work on the topics which are of interest to that laboratory). Those faculty who direct their own laboratories are usually not paid by the institution at which they are employed. Instead, they are almost always expected to obtain their own research funding, from which they pay their entire salaries. This funding also covers research expenses, the salaries of all laboratory personnel, and institutional overhead. These faculty usually do not teach at all, or they have a very minimal (and voluntary) teaching load.

Those faculty who work in someone else's laboratory may or may not be expected to obtain their own funding, or they may simply assist the principal investigator in obtaining funding. In this latter case, the non-tenure track faculty member's salary would be paid from the principal investigator's money.

Laboratory Technicians

Laboratory technicians can work in several different settings within academia. They can work in research or teaching laboratories. Additionally, they can work in clinical and diagnostic laboratories, in those institutions which are affiliated with hospitals and/or veterinary care facilities. Some institutions may also have gardens, arboretums, and other displays or nature preserves which require maintenance and attention.

Clinical Study Coordinators

University-based medical clinics and hospitals which conduct research have clinical study coordinators who are responsible for coordinating these clinical research efforts. These clinical study coordinators perform many of the tasks of the clinical trials such as subject recruitment and making sure that the study is being conducted and data reported according to the research protocol of the sponsoring institution. Although the principal investigator is usually a licensed medical doctor, the clinical study coordinator is this medical doctor's "right-hand man" in this regard. General information on the conduct of clinical trials is presented more thoroughly in Chapter 2.

Administrative and Managerial Positions

Academic institutions, like all other institutions, have individuals who oversee the programs, directions, and operations of the institution overall. These individuals may manage personnel, programs, or departments within the institution.

Some of the types of programs which exist at many academic institutions include:

• Technology transfer. The process of technology transfer involves the patenting and licensing of academic inventions to other users. Technology transfer is also described in other chapters of this book.

• Library and information science. Many academic institutions have their own medical and/or biology libraries which cater to individuals in these professions. As such, some institutions seek librarians and other information scientists who are knowledgeable about biology and/or medicine, as well as in the administration and management of libraries or of computer resources. These individuals may be responsible for establishing centralized resources and making acquisitions which are likely to be helpful to the institution's biologists and medical workers.

• Programs to promote the scientific activities of the institution to the public (e.g., public relations). For example, some institutions produce newsletters and other sources of information about the institution's activities. In these types of institutions, scientists are needed for such activities as writing articles about the institution and its research, and for promoting the institution's activities in the various news media and in the community.

• Programs to promote relationships between academic institutions and industry or government. For instance, some institutions have liaisons who try to cultivate relationships with industry and the government and to form partnerships with them. Many universities have grants offices which help researchers with obtaining grant applications.

• Programs to promote the safe use and disposal of laboratory waste. Many institutions have inspectors and health physicists who inspect institutional facilities for safety and who are available to answer questions regarding safe use of these types of research materials.

• Career counseling programs.

• Programs to ensure the ethical and humane treatment of animals and human subjects during research.

There are also many other types of programs which are administered and managed within academic institutions. For example, some institutions have clinical research programs which are administered and managed by the university (rather than by the individual researcher). Virtually all of the departments within an academic institution have some managerial and administrative workload. The types of managers and administrators depend very much on the institution, its programs, and its philosophies about what aspects of its work actually need to be managed. Nevertheless, there is a diversity of career opportunities in these areas. If you can think of a need for an academic institution which is not being met, perhaps you can even talk the institution into creating a position for you!

Chapter 7
Contract and Service Organizations

For every task that's required
A contracting firm can be hired.
They basically lease
Their own expertise
They do any job that's desired.

To illustrate the concept of contract companies, let's pretend that you are rich.

Once you can get into this mode of thinking, then read on. Let's pretend that you have an important business transaction. The problem is, you live in New York, and the transaction is supposed to take place in Texas. You are supposed to be there 5 hours from now. If you are not physically there, you will lose your chance at this great business opportunity.

You pace around the room contemplating this problem. You scratch your head, roll your eyes, and make a few phone calls. You find out that unfortunately, there are no commercial flights which can get you there in time. So what are you going to do?

You then have an idea. You'll charter a plane.

You make a few more calls and soon locate a pilot who is willing to perform this service. Within a short period of time, you are airborne and heading toward your destination on time.

Scientific contract companies, laboratories, and other independent service organizations are very much like the chartered plane. They can be hired to provide a scientific service. Their clients can be individuals, industries, government organizations, academic institutions - anyone who needs their services and is willing to pay for them.

Figure 13
Some Types of Contract Companies

Contract Manufacturing and Production Companies	These organizations can serve the needs of clients who do not have the facilities, manpower, and/or inclination to perform tasks in-house. For instance, a laboratory which needs very large numbers of cells may not have the space, time, personnel, or special facilities required to grow these large batch cultures. A biotechnology company may want to try manufacturing a product in a special way, such as through the use of a transgenic animal. A pharmaceutical company may need to manufacture several different study medications for a comparative trial, some of which they do not normally manufacture.
Contract Research Organizations	These organizations can provide research support in virtually any topic area. Areas can include basic scientific research, preclinical pharmacology and toxicology research, environmental research, clinical research, manufacturing research, and so on. Contract and service organizations in this line of work may test compounds or agents, perform biostatistical analyses, conduct epidemiologic studies, or provide a full range of services from the implementation of the research to the development and analysis of the database and the preparation of reports and regulatory submissions.
Contract Marketing, Advertising, and Sales Organizations	Some contract companies offer services in the areas of marketing, advertising, and sales. For instance, a contract company may be able to offer expertise and services in the areas of market assessment and pharmacoeconomics. A contract or service organization may advise a client organization on how and where to advertise products. Some contract organizations actually sell products for clients.

(Continued on next page)

Figure 13 (continued)

Scientific and Technical Contract Personnel Organizations	These organizations specialize in providing qualified personnel to client organizations. This arrangement involves the "leasing" of an employee to a client organization. In these cases, the employee works for the contract or service organization but often reports directly to work at the client organization, working side-by-side with the regular employees of the organization. The employee is usually leased to the client organization because he or she has some special technical expertise that the client needs in-house for a limited time.
Other Scientific Testing Firms	Some contract and service organizations offer services in the areas of scientific testing for non-research purposes. For example, some contract companies can perform environmental assessments of manufacturing facilities and products to assure compliance with regulations of EPA, OSHA, and/or FDA. Some companies may perform quality control tests on products such as foods or other biotechnology or pharmaceutical products. Tests which require sophisticated techniques and/or expensive equipment are often prime candidates to be outsourced to contract organizations by clients.
Other Scientific Consulting and Contracting Firms	Scientific consulting and contract firms (or individuals) can sell their expertise in many different scientific specialty areas, ranging from financial or regulatory advice to the installation of a new piece of laboratory or computer equipment or the implementation of a new laboratory technique. The possibilities for consulting work are extremely varied, and virtually anything is possible.

Career Opportunities in Contract Organizations

Contract companies can be hired to support the work of industry, government, or academia. They can pop up to fill any niche for which there's a market. Some examples of the types of organizations which currently exist are presented in Figure 13. Of course, there are also many more possibilities and opportunities than those which are listed here - you'll have to think about what services are needed in your area of expertise.

The concept behind contract and service organizations is that they can perform tasks more cheaply, conveniently, quickly, or with better quality than the client organization itself. However, the client wants the task to be performed according to its own specifications and standard operating procedures. Thus, the contract or service organization often must function much like an extension of the client organization itself.

With that in mind, and since contract and service organizations can support any or all of the different 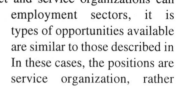 employment sectors, it is probably not surprising that the types of opportunities available with these support organizations are similar to those described in the earlier chapters of this book. In these cases, the positions are found within the contract or service organization, rather than in the sponsoring organization.

In any event, examples of the types of positions you can find within contract and service organizations include:

• *Research Scientists* for all types of topics. For instance, contract companies may work on developing state-of-the-art manufacturing or testing procedures to increase their business. Others may be hired specifically to conduct basic or applied research on a topic area picked by the client. For example, a pharmaceutical company or government agency may hire a contract company to perform preclinical or clinical research. In these cases, contract companies can have a wide range of positions associated with the conduct of the research (such as laboratory workers, scientists, clinical study monitors, positions in patient care, database managers, drug safety specialists, epidemiologists, and so on).

• *Regulatory Specialists and Consultants.* Many scientific contract companies serve organizations which are regulated by the U.S. or local governments (for example, environmental, agricultural, and pharmaceutical firms, and academic institutions conducting research). Sometimes these clients seek regulatory advice as part of the contractual arrangement because they do not know the regulations which govern their operations. At other times, the clients know the regulations, but the contract company must ensure that all work provided to these clients does indeed adhere to the applicable regulations. In either case, the contract company must have regulatory knowledge and expertise in order to provide good service.

• *Other Scientific Consultants* who can offer opinions, assessments, knowledge, and advice on any number of different scientific areas. While some consultants may offer regulatory advice, other consultants may offer any number of different services from financial advice for scientific companies to the implementation of a new laboratory technique or computer system. They can help pharmaceutical and biotechnology companies with development plans for their agents or can help them to design their preclinical or clinical research studies. They can advise a company on regulatory issues such as how to prepare for a regulatory audit. Scientific consultants can either work on their own (that is, "freelance") or can be affiliated with a contract organization. They can work on any topic area or specialty for which there is a market.

• *Scientific Equipment Maintenance People* who can repair and maintain certain types of scientific equipment. These people may also inspect equipment on a regular basis to ensure that it is functioning properly and according to any regulatory guidelines which may apply.

• *Laboratory Technicians* to do procedures such as manufacturing production, growth of cell cultures, quality control procedures, environmental assessments, analysis of clinical specimens, and so on.

• *Auditors and Inspectors* to perform procedures as they relate to a quality control or quality assurance program (described more thoroughly in Chapter 2), and/or as they relate to compliance with governmental regulations (for example, the regulations of the Occupational Safety and Health Administration or the Environmental Protection Agency). The exact nature of the work (benchwork, fieldwork, and/or deskwork) depends on the type of audit or inspection which is being performed, and for what type of organization.

• *Computer, Information, and Database Specialists* to manage databases and information for research and other programs (for instance, DNA sequences, preclinical or clinical data, epidemiologic data, pharmacoeconomic information, and so on). These databases must be created in a manner which is amenable to later analysis and/or reporting (which may also be performed by the contract company).

• *Biostatisticians* to advise on matters of data analysis, sample sizes, and other relevant statistical issues.

• *Scientific Writers and Editors* to write regulatory submissions, product monographs, scientific or technical advertising, technical manuals, regulations, informational tools, and so on.

• *Scientific and Technical Recruiters* to identify and recruit personnel who possess certain skills and can be "leased" to client organizations. Placement of individuals into appropriate employment situations requires a knowledge of the skills and scientific expertise that is required to perform certain types of jobs.

• *Patent and Other Technical Lawyers* who can provide legal expertise to their clients in topic areas such as patent law, technology transfer, forensics, scientific and medical ethics, regulatory law, and the execution of technical contracts between contract companies and their clients.

•   *Contract Salespeople* to describe and sell technologies such as laboratory or medical equipment, medical devices, diagnostic tests, pharmaceutical and biological agents, or other laboratory supplies and reagents, to their target markets, either in person, by mail, or over the telephone. Contract salespeople in the pharmaceutical industry are often called "detail" people (i.e., they "detail" products to physicians - please see Chapter 2 for more information).

•   *Contract Marketing People* who conduct market studies and develop marketing programs for organizations on a contractual basis. This work may involve the design and conduct of Phase IV studies for the pharmaceutical or biotechnology industries. It may also involve research of market size, the costs, benefits, and liabilities of competing products, and other research which can enable a company to make decisions about product pricing.

•   *Administrators and Managers* to manage and administer projects, programs, personnel, contracts between the company and its subcontractors, or between the company and its clients, and money. These individuals often liaise with clients, as well as make decisions about internal company affairs.

•   *Business Developers, Salespeople, and Marketers* who identify and target potential clients. Thus, these companies need individuals who are familiar with the scientific niche they are trying to fill, and who may be familiar with potential clients. These individuals may also be able to assess the relative sizes of different potential markets, and to develop business strategies using this information. After identifying potential clients, these individuals try to call or meet with them to explain the services the company can offer, in an attempt to solicit their business.

It is noteworthy that contract companies display a wide variation in the scope of services offered. For instance, a contract research organization supporting the pharmaceutical industry may offer a full range of services from preclinical research to regulatory submission to postmarketing. Another contract research organization may only offer services in the area of clinical study monitoring. Obviously, the types of jobs available within any particular company depends on the scope of services that company wishes to offer its clients.

Employment in a contract organization can be an excellent way to gain the experience you may need to break into a particular field or type of work. Because of the contractual nature of the work, you will be exposed to many different clients and projects, and you will see many different philosophies and approaches. It is therefore a great way to gain a broad perspective about a particular industry or line of work.

Chapter 8
Other Organizations in the Private Sector

Hospitals, gardens, museums, and zoos,
So many options from which you can choose.
They're not pharmaceutical, not biotech,
But still they can pay you a livable check.

In writing this book, I found that certain employers just didn't seem to fit into any one category. They definitely weren't pharmaceutical or biotechnology companies. They also weren't government or academia. They weren't really contract research organizations, laboratories, and other independent service providers, either. Finally, I decided to put them into a category all their own.

You may have already noticed that some of these types of employers and positions are also mentioned in Chapter 7 of this book. The reason for this is that some organizations may have more than one purpose. For example, a veterinary clinic could have a contract with an industrial organization (for example, an agricultural firm), and could also also work with people's pets. Further, sometimes the distinction can be blurry between a contract organization which supports the development and manufacture of products, and an organization which offers other types of products or services.

Figure 14
Examples of Potential Employers of Scientists

Manufacturers and Suppliers of Scientific Equipment, Reagents, and Software

Software is needed in such areas as bioinformatics, as well as for such uses as organization of literature references and analysis of laboratory or other scientific data. The advancement of science and medicine is also facilitated by the development of new and improved scientific equipment and technical capabilities.

Scientific Publishing Firms

Scientists are needed to solicit, edit, and write works for publication, either as books or news articles. Scientists are also needed to make decisions regarding the publications, including the identification of hot topic areas and the overall focus of the publishing firm.

Scientific Law Firms

Scientists are needed in the areas of patent law, technology transfer, scientific ethics, medical malpractice, and forensics, to name a few. Scientists may also have law degrees, although they are not always necessary. Scientists are needed to research cases. Scientists can also provide expert testimony in cases involving scientific evidence.

Venture Capital and Investment Firms

Scientists can couple their scientific skills with financial knowledge to advise "business incubators" (which provide start-up space and money to new scientific businesses), investors, and other venture capitalists who are interested in investing in technical and/or biological organizations.

Human and Animal Care Facilities

Individuals with a scientific background can work in physicians' offices, hospitals, veterinarians' offices, animal hospitals, and similar types of facilities. These positions can involve direct patient care as a medical doctor, veterinarian, or an assistant. They can also involve work in a specialized scientific area (for instance, genetic or fertility counseling, or assisting with procedures for in vitro fertilization). Some medical clinics and hospitals also conduct clinical research. In these facilities, there are often opportunities for clinical study coordinators, who are responsible for recruiting patients into their clinical studies and assuring adherence to the study protocols.

(continued next page)

Figure 14 (continued)

Health Maintenance Organizations

Scientists are employed by health maintenance organizations (HMOs) to understand and interpret medical issues, such as the prevalence of certain medical conditions within society and its sub-groups. These scientists may also analyze risk factors for disease and the effectiveness of various treatments. This information is used by HMOs for pricing, and also to make policies regarding which medical treatments they will accept (and reim-burse).

Scientific Professional Societies, Private Foundations, and Special Interest Groups

Scientists can work within these groups as advocates of particular scientific policies (either to the government, as lobbyists, to the public at large, or to the scientific community). Additionally, scientists may be involved with other activities such as the planning and implementation of scientific meetings, training courses and seminars, career placement services, fundraising, and grant adminis-tration.

Scientific and Technical Recruiting Firms, Employment Agencies, and Other Types of Headhunters

These firms can employ scientists who understand the types of skills necessary to perform certain tasks. Scientists who work in these firms must also have connections within the scientific community. Scientists who work in this capacity may locate workers who are then hired by client organizations on a temporary or "permanent" basis. This arrangement is unlike the contract employment arrangement described in Chapter 7, since the arrangement in Chapter 7 assumes that the employee ultimately works for the contract organization and not for the client company.

Zoos, Nature Preserves, Museums, Aquariums, Sanctuaries, Botanical Gardens, Circuses, and Other Exhibitors of Nature

Scientists are needed to select, organize, and present displays and exhibits (a process which is often referred to as "interpretation"). In addition, scientists may administer their operations of the organization, give educational talks, train animals for performances, conduct performances, and take care of the animals and/or plants on a day-to-day basis. Some organizations have full-time veterinarians, botanists, and other specialists who have training in the care and maintenance of plant and animal life.

(continued next page)

Figure 14 (continued)

Non-Pharmaceutical, Non-Biotechnological Industrial Organizations	Scientists may work for companies which have products that are completely unrelated to biology. For instance, scientists may study a biological process because it relates to the manufacture of a non-biological end product (such as a plastic which can be biodegraded, or the manufacture of snow by using a product which is made by a microorganism). Scientists are also needed by manufacturers of foods and other agricultural products, to perform microbial, biochemical, and/or any other battery of tests which are required or desired by the industry, and sometimes to study packaging issues as well. Scientists may also be hired by many different types of companies (with either biological or non-biological products) to perform environmental or workplace tests to assure compliance with the regulations of governmental agencies such as the Environmental Protection Agency and/or the Occupational Health and Safety Administration.
Private Testing Laboratories	Clinical and diagnostic laboratories are often affiliated with physicians' offices, veterinarians' offices, hospitals, pharmaceutical and biotechnology companies, government institutions, and medical and veterinary schools. Forensic and environmental testing laboratories can also be a source of employment opportunities for scientists.
Other	There are many other types of employers not mentioned here which can be a source of employment opportunities for scientists. For instance, there are opportunities in scientific advertising, scientific illustration, and more. It is impossible to create a comprehensive list of every type of firm which may employ scientists; instead, you are encouraged to think about the types of work that you like, and how they might translate into employment opportunities for you.
Entrepreneurial Opportunities	If you have an idea for a new scientific product or service and you think there's a market, you may wish to research it, start your own company, and go for it!

Types of Positions

Here are some of the types of positions for which technical expertise is helpful or required, and which are available to you with some of the different types of employers mentioned in this chapter:

• *Research Scientists.* Research scientists conduct research on many different topics and for many different purposes. They can work for HMOs, private foundations and special interest groups, private laboratories, and manufacturers of scientific reagents, diagnostic assays, and equipment. In some cases, the research may involve benchwork, and in some cases, it may involve analysis analysis of clinical or epidemiological data. This information may be used to further develop the company's products, or it may be used in decisionmaking regarding the types of programs and policies the organization will support, and perhaps try to promote (and at what cost).

• *Developers of Scientific Software and Equipment.* These creative individuals develop many of the technologies and software which are used in research and diagnostic laboratories, in physicians' and veterinarians' offices, and in the field (for instance, at a crime scene or in the environment). For instance, these individuals may develop computer programs which analyze and compare DNA sequences (often called "bioinformatics"), do molecular modeling, and analyze and graph data, or they may develop new technologies such as high performance liquid chromatography or flow cytometry. Although these activities might be considered research, their activities are sufficiently different that they are included here in their own category.

• *Laboratory Technicians.* Laboratory technicians are needed to perform diagnostic procedures, environmental tests, forensic tests, and environmental tests, for organizations who offer these services. In addition, laboratory technicians can be used to support the research and manufacturing activities of organizations, for example, by performing quality control measures, trials of new laboratory equipment, by maintaining cell culture and/or animal facilities, or by performing any other experimental procedures required by the employer. Laboratory workers can play an influential role in conducting laboratory tests for audits and inspections of products, the outside environment, or the environment within the workplace.

• *Quality Control and Quality Assurance Auditors and Inspectors.* These individuals perform quality control tests or audits as described in the other chapters of this book. Obviously, the nature of this work depends very much on the types of products a particular organization produces, and the regulatory agencies which govern it.

• *Biostatisticians.* Biostatisticians are employable within any institution which requires statistical input. For instance, HMOs may wish to analyze data statistically to decide on the types of treatments and policies they will support. Other organizations which conduct research and/or analyze databases may have the need for biostatistical input as well.

• *Computer Programmers and Database Managers.* Many institutions (such as research organizations, HMOs and other special interest groups) collect, develop, maintain, and analyze large databases of scientific information for various purposes. For instance, HMOs and special interest groups may develop and analyze these databases in order to determine their policies and positions on various issues. These institutions need individuals who have a good understanding of how to create and maintain accurate and clean databases of this information, in a way in which it can be meaningfully utilized and/or analyzed.

• *Technical Support and Training.* Organizations which manufacture products such as software, reagents, and scientific equipment usually offer technical support for the use of their products. This is particularly true of highly technical products which require some technical expertise in order to use them. Technical support and training can be offered over the phone or in person, either at the customer's location or at the vendor's.

• *Scientific and Technical Recruiters* to identify and recruit personnel for scientific and technical industries, either on a temporary or "permanent" basis. Placement of individuals into appropriate employment situations requires a knowledge of the skills and scientific expertise that is required to perform certain types of jobs.

• *Product Marketing.* Organizations often make a science out of maximizing their revenue from sales of their products and/or services. Along these lines, these organizations may hire scientists to conduct market studies, to determine the potential market size for the item(s) they are promoting, and to determine the price that the marketplace is willing to pay. These types of studies, which may be conducted in conjunction with other studies (such as epidemiologic studies) may also help the organization to look for ways to increase market size or conduct studies to try to determine the most effective way(s) of promoting and advertising.

•   *Product Sales.* Technical salespeople are the movers and shakers who actually pursue the target market and try to generate sales of the product or service. Sales activities may involve traveling to different locations to promote the product, telemarketing, writing letters and sending promotional flyers to customers. Salespeople may use any number of different promotional gimmicks and sales tactics.

•   *Lobbyists and Public Policy Analysts.* Some industrial organizations and special interest groups hire scientists to analyze and/or promote technical policies as they relate to their industries. These people identify important policy issues and gather evidence to support or reject certain types of policies. Since many organizations are interested in influencing public policy, these organizations may also have lobbyists and people in public relations, who try to influence politicians and/or the public.

•   *Venture Capital and Investment Advisors.* Scientists who have an interest in investments and venture capital can work as financial advisors for business incubators, for stockbrokers, and for businesses. These scientists may try to raise venture capital for companies, or they may work for business incubators who are looking to invest in scientific companies. Some scientists perform financial analyses on different companies and communicate the results via publication or some other format.

•   *Administrators and Managers.* Like the administrators and managers described in other chapters of this book, scientists in these positions may be responsible for administering their projects and programs (for example, grant and research programs, policy promotional activities, meeting planning, or other activities). These individuals, or a separate group of individuals, also keep track of monetary and budgetary issues, manage and allocate personnel, and decide on the overall directions and focus of the organization. Although the responsibilities of administrators and managers are quite varied, depending on the missions and purposes of the organizations in which they are employed and how the organization is structured, there generally are numerous opportunities for scientists along this career path, just because of the large number of different programs and projects which must be managed and administered.

•   *Writers, Editors, Publishers, and Other Scientific Communicators.* Scientists are hired by many of the institutions, organizations, and special interest groups mentioned above, to convey information or to promote the institution or its products and services. Audiences for these materials may be scientists, the general public, shareholders or potential investors, politicians, patients, and/or potential customers. Any or all of the different communication media may be used, including informational pamphlets, television, radio, newspaper, public lectures, and so on. Scientists are also hired by publishing firms to submit, solicit, and edit scientific work. Although in some cases, the job market can be very competitive and limited to scientists with considerable experience (particularly with journals which publish peer-reviewed research articles), many publishing firms offer opportunities for less experienced scientists as well.

•   *Scientific Lawyers and Legal Assistants.*   Scientists can use their scientific knowledge to work as a lawyer or a legal assistant. Organizations which may need technical lawyers include law firms, special interest groups, and numerous industrial companies. Scientific training is especially useful in the areas of patent law, some cases of criminal law (especially for cases involving forensic scientific evidence), scientific ethics, medical malpractice, regulatory law, environmental law, and scientific policy.

•   *Organizers of Meetings and Scientific Programs.* Biologically-oriented groups often organize meetings of a technical nature in order to promote the advancement of certain scientific or technical topics. These meetings are often coordinated by individuals who work full-time within a special interest group or professional society, and who are aware of the scientific topics and individuals which are most likely to be of interest to that particular scientific community.

•   *Patient Care.* Positions are available for medical doctors, nurses, and physician assistants. However, there are many other positions in patient care as well. For example, there are physical therapists, genetics and fertility counselors, pharmacists, exercise physiologists, and nutritionists, just to name a few. These types of positions can be found in private practice, in hospitals, spas, pharmacies, and/or on an independent, "freelance" basis.

•   *Animal and/or Plant Care.* Zoos, institutional or private gardens, nature preserves, and other animal and/or plant facilities can provide a number of employment opportunities for biological scientists to take care of their facilities and displays. For instance, veterinarians and veterinary assistants can care for zoo animals. Botanists and landscape architects can design and care for private or institutional gardens or displays. People with a knowledge of plants or animals can design exhibits and shows for museums, and animal trainers can train animals to perform in shows. Ecologists can work in nature preserves and at other institutions which display, study, and/or have a focus on nature.

•   *Other.* There are many, many opportunities for scientists to work within many different types of organizations. You are limited only by your creativity, and how far you wish to deviate from your original scientific training.

•   *Self-Employment.* In reading through the employers and positions that I have presented, you may have thought about what it would be like to work for yourself. Indeed, there are many ways to do so. Perhaps you have expertise, skills, or technical knowledge which you could lease to someone else. Alternatively or additionally, perhaps you have an idea for a product of your own!

Whether you will wish to work for someone else versus on your own will probably depend on many factors. Do you have sufficient expertise and resources to go out on your own? Will your client organizations or customers hire you as an individual or buy your product, or do you need the backing of a larger, more established company? Finally, how adventuresome are you?

In Closing
You may have noticed in reading this section that there is not always a clear delineation between the responsibilities of the different types of positions described herein. This is because there are so many different types of organizations which exist for various missions and purposes. Obviously, the purpose of the organization determines the types of work it performs. Additionally, as described in Chapter 9, individual organizations can be organized differently with respect to how they divide the workload among their workers. Outsourcing can also have a profound effect on the types of jobs which are available in the organization (the jobs within organizations which outsource most of their work tend to be more managerial or administrative in nature than hands-on).

Chapter 9
Organizational Structure and Job Titles: What Do They Mean?

With many a task to be done,
An organization is run
To do what it needs
So that it succeeds
And functions as if it were one.

The previous chapters of this book have discussed some of the job opportunities that may be available to you with various employers. After reading them, however, you may be confused about one issue. How in the heck are you supposed to understand the job titles in the "real world"? Often, they seem to be written in some unknown foreign language which makes it impossible to understand what the job is really about!

The purpose of this chapter is to help you to make sense of job titles and job descriptions. After reading this chapter, you may wish to go back and re-read some of the previous chapters with a new perspective. If you do, you will have a much better understanding of how the working world works, and how different positions and titles relate to certain types of work.

Once you finish reading this chapter, much of the information will probably seem quite obvious to you. In fact, it *is* obvious - *once you think about it*! However, I know that I didn't think about it when I first ventured out on my job search. The purpose of this chapter is so that *you* do!

Organizational Structure

Job titles and descriptions are easier to understand in the context of an overall organizational structure.

To start, let's think about a business we're all familiar with. It's called a household. For the purpose of this chapter, we will call it a "domestic business unit" (DBU).

In fact, a household (DBU) is very much like a business. Each must find some way to generate revenue. Additionally, each must perform other chores which are necessary for continued functioning of the unit.

Let's now examine the structure of two identically-sized DBUs (Fig.15).

Figure 15
Two Domestic Business Units (DBUs)

DBU Member	DBU #1	DBU #2
Husband	• Earns income • Pays bills/manages money • Does income taxes • Maintains outside of house • Maintains cars • Hires outside help (see below)	• Earns income • Shops • Cooks • Does dishes • Takes out garbage • Helps raise children • Hires outside help (see below)
Wife	• Shops • Cooks • Cleans • Does laundry • Raises children	• Earns income • Shops • Cooks • Does dishes • Does laundry • Pays bills/manages money • Helps raise children • Hires outside help (see below)
Child #1	• Does yardwork • Takes out garbage • Cleans room • Studies	• Studies • Cleans room
Child #2	• Does dishes • Vacuums whole house • Cleans room • Studies	• Studies • Cleans room
Outside Contractor	• Does difficult car maintenance • House painting and plumbing	• Yardwork, lawn mowing, snow shoveling • Cooking (this DBU often eats out) • All house and car maintenance • Does income tax • Helps raise children • Cleans

Upon examination of these two DBUs, one thing is inherently obvious. These two DBUs have entirely different approaches to running their units.

Why?

The chiefs of each of these DBUs (husband and/or wife) want to divide the tasks in a way which works best for them. This concept may be different in different households. Oftentimes this division of labor is also based on the aptitude, availability, and willingness of the members of the DBU to perform certain tasks.

For example, perhaps the husband in DBU #1 likes to do car and house maintenance, and the wife likes to cook and wants to be home with their children. The chiefs in this DBU also believe that children should be raised to have some chores around the DBU. Thus, for DBU #1, their organization of chores may get the work done in a way that is agreeable to each of the DBU's chief officers.

In contrast, the wife in DBU #2 may want to work (or the husband may want her to). She may earn considerably more income by working outside the home and hiring chores to outside contractors than by staying home and performing more household chores herself. This DBU may also believe that the children should spend their time learning and playing, not doing routine household chores. Thus, DBU #2 has a completely different organization of chores than DBU #1.

So, you might say, how does this relate to the working world? The points I'd like to make are this:

•   Businesses, even of the same type, can distribute their tasks differently from one another. The organization of any one particular business can depend on the schedules, aptitudes, attitudes and/or preferences of its members.

•   Not all businesses perform the same exact set of tasks. Northern DBUs have to shovel snow, whereas southern DBUs don't. Southern DBUs have termite and pest problems, whereas northern DBUs generally don't. Some DBUs maintain their own housing facilities, whereas others operate from apartments. Some have cars and some don't. Some DBUs have higher standards of cleanliness than others. Thus, the tasks that are done by any given DBU are influenced in part by its income, its geographical location, its philosophies about what constitutes a pleasant and successful living environment, and countless other factors.

The working world is like a number of different households. Each organization has tasks to accomplish. Although their missions may be different, each organization is faced with the same problem of dividing its tasks among its workers so that all the work gets done.

Now do you understand why you are confused about organizational structure and job titles? It's because there's no one standard way of doing things. You're stuck trying to figure out how each organization is structured on an individual basis.

So I wish I could give you an overview of The Official Organizational Structure. Unfortunately, though, I can't.

Job Titles

Now let's take things a step further. Let's assign job titles to the members of the DBU.

Here are some proposed position titles:

Figure 16
Two Domestic Business Units - Position Titles

DBU Member	Position Title - DBU #1	Position Title - DBU #2
Husband	President and Chief Financial Officer	Chief Executive Officer
Wife	Chief Operating Officer	President
Child #1	Senior Apprentice	Assistant 1
Child #2	Junior Apprentice	Assistant 2

What strikes you about these titles?

First, you may notice that the position titles have a certain psychological effect on you. They encourage you to focus on certain roles of the worker (either real or perceived). Secondly, you may notice that the job title "President" is used in both DBUs. However, the President in DBU #1 has a different set of responsibilities than the President in DBU #2.

Obviously, then, you can't always tell the responsibilities of a particular job by its title. To give you a "real world" example from the pharmaceutical industry, a Clinical Research Scientist can be either a medical writer or a person involved with the monitoring of clinical trials. These are, in fact, two very different jobs, but you'd never guess that from the job title.

To complicate things a little further, let's consider the fact that the same person, in the same overall job, can be referred to by multiple titles. For instance, the husband in DBU #1 can be called husband, father, dad, papa, chief, president, and so on.

So what's the point? YOU CANNOT JUDGE A POSITION BY ITS TITLE!

So what are you supposed to do?

What You Can Do To Understand It All

There is one simple way to make sense of organizational structure and job titles. You can ask your potential employers to define them when you are applying for positions!

Job Descriptions

Job descriptions are meant to clarify the responsibilities of the position.

So ask for them! Of course, you will not necessarily understand position descriptions, either, unless you have a good general understanding of the organization and/or type of work for which you are applying.

Organizational Charts

Many organizations have organizational charts. These diagrams can show you exactly where your position fits in to the overall operations of your division, department, or overall organization. So ask for an organizational chart - the employer will perceive you as being interested in the organization overall and may even think favorably of you as a potential team player.

Other Information

Get the employer to talk about the organization and position as much as you can. Ask all sorts of questions - anything you can think of - about the organization and position. By doing so, you may find out additional aspects about your job that you may not have thought of, or which there would be no way of ascertaining from the job description or job title. For instance, you may wish to learn about the history of the organization, since an organization's history will often help to define its structure and division of labor.

A Note on Organizational Size

Organizational size can have a very strong influence on the types and flavors of jobs within a particular organization, as well as how a particular job fits into the overall functioning of the organization.

For example, large organizations can have people who are very specialized in one area. This would be similar to a large family having one person who is a very good cook. This may be the only chore that this person does in the family, but this person does it very well, and for a lot of people. In smaller organizations, each individual often has to become more of a jack-of-all-trades, just because there aren't enough people to assign a separate person to each chore.

Of course, large organizations are not always structured that way. Some large organizations, such as large universities, may have separate departments, laboratories, or other business units, each of which functions somewhat independently of each other.

Nonetheless, it is a wise idea to ask about the size of an organization and how it affects the jobs within it. Oftentimes you will find that large organizations are very much like large families. The larger their size, the more inertia there is to overcome, and the more there is to organize in order to get anything done. To cope with this problem, large organizations often form organizational layers, each of which reports to the next level up. While this can increase the amount of bureaucracy in an organization, it can also result in a large number of resources and skilled professionals with whom you can train and from whom you can learn!

In Summary

The previous chapters of this book are intended to describe the types of work performed by the organizations and employers described in them. However, because job titles and organizational structure differ so much among organizations, the position descriptions are *not* universal position descriptions among these organizations - they are merely *examples* of the types of positions you might be able to find in those types of organizations. ***It is up to you to think globally*** about the types of work performed by these organizations and to ask exactly how the work is divided by the employers which interest you.

The pharmaceutical industry is an excellent example of how an organization can accomplish and coordinate a multitude of scientific tasks. You are therefore encouraged to read that chapter, even if you think that you are not interested in that type of work. Each organization within that industry functions a little bit differently, depending on whether each person performs one task versus a multitude within the organization. Pharmaceutical organizations often also differ from one another with regard to job titles for the same job.

In any event, asking the right types of questions during your job search can save you a lot of headache, heartache, and misunderstanding about the job to which you think you are applying. It is NOT naive to ask what a job title means, and to ask the employer exactly what is expected of you in a particular position. After all, how could you possibly be expected to know the details of each and every position with each and every employer, given the diverse array of organizational structures and job titles and the lack of homogeneity among these different employers? At least if you ask, you may be saving yourself from the unfortunate position of accepting a job without really understanding (or liking) what is expected of you!

**Don't Get Lost -
Ask for Directions!**

Part III
If Opportunity Doesn't Come Knocking, How Can I Find It?

Chapter 10
How to Identify Job Opportunities

The end of your schooling is near,
The time for your job search is here.
But with all that college,
You've found that your knowledge
Won't help you to find a career.

In many ways, searching for a career is like playing a game of hockey. The puck is always moving, and players are constantly changing directions as they re-evaluate their positions and their strategies, depending on the behavior of the other players. Although their motion may at times seem quite arbitrary, random, and even unproductive, good players are always focused on their goals - which include both their own and those of the competition.

Like an intense game of hockey, the types of positions in the world of work are always changing. So how do you know what's hot and what's not in your field, and what the specific opportunities are in your field at any given moment? Somehow you need to know the positions and strategies of certain key players in relation to the puck. So how do you find out?

You may have noticed that the previous chapters of this book do not provide this type of information. Those chapters are designed to help you to figure out what position you might wish to play. However, they still do not give you enough information to fully understand and appreciate the different opportunities which are available to you, nor do they provide sufficient information about the scope of duties and responsibilities that each position entails.

The reason for this is discussed more fully in Chapter 9. That chapter emphasizes the fact that organizational structure, as well as philosophy, size, mission, and scope, differs from organization to organization. Like hockey clubs, each organization uses an individualized strategy to get its players to work together cohesively. Thus, each organization is somewhat unique with regard to the exact nature of the positions available within it.

Nevertheless, different organizations, like different hockey clubs, are often structured somewhat similarly to one another. These similarities have been the basis for the generic information on positions in the previous chapters of this book.

In any event, you're still going to have to do your own research in order to learn more about the opportunities that are available specifically to you. The purpose of this chapter is to help you to do this. You may wish to refer also to Chapter 15, which contains detailed descriptions of some specific positions. These descriptions can help you not only to learn more about the specific types of positions which are described in that chapter, but also to help you to think about the types of questions you may wish to ask in your own exploration. The Appendices also contain listings of resources which could help you in your journey.

So what are some of the techniques that you can use to begin your own research into the job opportunities which are best for you?

General Sources of Job Information

Personal Contacts

What better way is there to get information than to talk with other people? You can learn a tremendous amount of information in this way. You can listen, ask questions, and truly get a sense of what a position entails. You can gather a lot of information in a very short time.

That's great when you know someone who works in your field of interest. But what about when you don't? Luckily, there are many ways to go about meeting people who may be able to help you. One of the most favorite techniques for doing this is called "networking".

"Networking" is simply a fancy name for meeting people through people. People have extensive networks of contacts. You never know who knows who. So talk to lots of people - friends, family, alumni, classmates, faculty, co-workers, and people at professional meetings. Ask for "informational interviews". You never know who may be able to help you, so meet and talk to as many people as you can. Schmooze.

Professional societies

Professional societies can be great sources of information about a particular profession. Besides being a great place to make personal contacts, professional societies often compile and publish information about types of jobs, job markets and employment outlooks, and salaries in their professions. Additionally, professional societies often compile information about the type of education you will need to break into the profession, what types of schools offer these types of educational programs, resources and points of contact which are likely to help you in your job search, and directories of relevant employers or sources of information. A few good examples of professional societies which do this are the Biotechnology Industry Organization (BIO), which publishes a pamphlet on careers in biotechnology, and the National Society of Genetic Counselors, which publishes all sorts of statistical and other information about careers in genetic counseling. You are urged to refer to these published sources of information for accurate information about careers in the fields of interest to you, since they are updated more often and are likely to provide more accurate salary, education, and employment information than the sketchy information which is provided in Appendix A of this book.

Many professional societies also publish job advertisements and/or have placement services for people looking for jobs in certain professions. A listing of some professional societies which may be of interest to you is included in the back of this book. Further, since this list is by no means comprehensive, you may wish to enlist the help of a college or university professor and/or a librarian to help you identify additional professional societies which may be of interest to you.

School Career Placement Services and Alumni Offices

Many people are not aware that their colleges and universities have career placement offices which could help them with their job search. This is true for alumni as well as for active students - many of these schools offer their services to both of these groups.

A college or university placement office may be able to help you in several different ways. Most centers maintain a library which offers books and information on different types of careers. Some schools have also started to compile statistics about the career paths of their graduating students. Many placement centers have career counselors who can help you to plan your career-seeking and job searching activities.

Some school placement centers and/or alumni offices also maintain databases of alumni who have volunteered to talk with interested students and/or alumni about their professions. Additionally, some schools support "shadowing" programs in which a student can accompany an alumnus to his/her workplace to see the day-to-day activities which are involved in the pursuit of a particular profession.

School placement services can also help to put you in touch with employers, and to help you with your resume-writing and interviewing skills. These topics are discussed in the next chapter.

Professional publications and newsletters

Every profession has its fair share of professional publications such as journals, newsletters, and other sources of news and information.   These publications can be a great way to learn what's hot and what's not, and to learn about the current issues in your career or profession. You can often learn about the major employers in the profession. Sometimes you can even identify potential employers even by looking at the organizations who advertise their services in these publications!

A few professional publications, such as *Science, Genetic Engineering News,* and *The Scientist,* contain numerous articles on trends in science in general. *Nature* also has a series of publications (such as *Nature Biotechnology, Nature Genetics, Nature Medicine,* and *Nature Structural Biology*), which contain information about particular topic areas of science. In addition to basic science, these types of publications cover topics such as employment trends, venture capital (as it relates to biotechnology and pharmaceutical companies), grant funding, education, government agencies and policies, news about specific pharmaceutical and biotechnology companies, Internet sites, and numerous other topics of general scientific interest. For this reason, these types of publications are excellent for the job seeker who is looking to learn more about a particular topic area. *Genetic Engineering News,* and *The Scientist* are available free of charge or for a very small fee. The other publications are also available for a slightly higher fee.

There are numerous other publications which could help you to learn more about trends in your specific area of interest. Again, you may wish to consult a professional society for recommendations, or ask a librarian or a professor. Some of these publications are also listed in the Appendix and are discussed further in the following sections of this chapter.

Job Advertisements

Job advertisements are useful not only for the purpose of answering them, but also for obtaining all sorts of information about jobs. For starters, job advertisements can help you to learn about the types of jobs that exist within a certain profession, as well as the amount and type(s) of education that employers are looking for. Job advertisements can also help you to see what other types of credentials may be required for the job, as well as the salary range that you might be able to expect in a given type of job.

If you use job advertisements as a way of obtaining information, you may wish to be aware of one thing - employers often ask for more credentials for a particular type of position than they are likely to get. Thus, while these advertisements can provide a guideline of what is desirable to an employer, you can oftentimes land a similar position with less experience or education than what is requested in the job advertisement. Thus, you may wish to do more research before getting too discouraged or immediately jumping into an educational program to increase your credentials for a particular type of position. On the other hand, if you repeatedly see a requirement for a particular credential in virtually all of the job advertisements for a particular type of position, chances are that that credential is required for that type of position.

If you are interested in using job advertisements as a source of information, they can be found in numerous ways. They can be found in professional journals and publications of professional societies, in regular newspapers, on bulletin boards in university career centers, at scientific meetings, and over the Internet. Obviously, the types of places in which you look will depend on the type of job that you seek. You may wish to ask a professor, a professional society, and/or a librarian for help in identifying places to look for job advertisements in your particular profession. A listing of some resources which may be of interest to you is also provided in the Appendix.

If you are planning on using regular newspaper advertisements as a source of information, there are a few strategies that can help you. You may wish to refer to the book <u>What Color Is Your Parachute</u>, by Richard Nelson Bolles (listed in Appendix C), for lists of newspapers you may be interested in obtaining. As discussed in that book, you may wish to choose those newspapers which have a wide circulation, since these newspapers are likely to have job advertisements for a wide variety of professions and geographic areas. Additionally, you may wish to look at those newspapers in geographic areas which have a high concentration of organizations in your profession, even if you are not planning on relocating to that area, since job advertisements from these newspapers can give you a great flavor of the types of positions which exist in a certain profession, and since these newspapers may also have job advertisements for positions in geographic areas outside of the area in which the newspaper is published.

The Internet

The Internet is obviously a tremendous source of information on jobs. However, you may have noticed that this information superhighway is often devoid of road signs. For this reason, the Internet is the focus of a separate chapter of this book (Chapter 12).

Local Chambers of Commerce

Local Chambers of Commerce may be able to help you identify companies or organizations in a particular geographic area. To locate the Chamber of Commerce for a particular region, you may wish ask at your local library (particularly a business library), or look in that area's phone book. The local Chamber of Commerce can provide you with all sorts of information about jobs, cost of living, and other topics. This information is often provided either free of charge or for a nominal fee. The regional phone book can also provide a wealth of information about a particular geographic area.

Doing Internships, Co-ops, Work-Study, or Otherwise Gaining Real Job Experience

One good way to find out about job opportunities is to work in the field in which you are interested. Even if you do not yet have the credentials to work in the actual job to which you aspire, oftentimes you can still work as some kind of an assistant to a person who does do the type of job in which you think you are interested.

There are many ways to gain actual work experience in your chosen field. Some possibilities include summer jobs, college work-study jobs (either during the school year or over the summer), college co-op programs, internships, and volunteer work. These types of work opportunities are available in all employment sectors including government, the private sector, and academia.

There are many ways to identify these types of opportunities. You may wish to talk to your college or university and find out what programs or positions they offer. You may also wish to contact organizations directly, since many of them do have internship programs (you may be able to talk them into creating one, even if they do not have a formal program!). For additional internship opportunities, you can also contact the National Society for Internships and Experiential Education, The Princeton Review (which publishes "America's Top Internships" and "The Internship Bible") and "Arco Internships: A Directory for Career-Finders" (Macmillan). Addresses for these organizations are given in the Appendix.

How to Identify Job Opportunities in Specific Fields or Economic Sectors

The purpose of these following sections is to identify some specific ways and resources for learning about certain career paths and employers. You may also want to refer to the Appendix for a more specific listing of resources in different professions. Librarians, personal contacts, professors, and/or professional societies in your field of interest may also be able to refer you to informational resources which may be useful to you.

Jobs within the Pharmaceutical/Biotechnology Industry

There are several sources of information to learn more about the development process for pharmaceutical and biotechnology products which are regulated by FDA. These include a series of books published by Parexel, which describe the development process for pharmaceutical and biotechno-  logical agents. There are also a series of meetings and videotapes on the FDA regulatory process which are sold by BioConferences International. Please see the Appendix for specific information on how to obtain more information about these resources.

The *Applied Clinical Trials Journal* is available free of charge and is a good source for learning about the issues which are facing clinical researchers in the pharmaceutical and biotechnology industries. This journal also compiles information on interesting websites for clinical researchers and occasionally provides profiles of certain types of jobs within the industry. As mentioned above, *Genetic Engineering News* and *The Scientist* also cover numerous topics in the pharmaceutical and biotechnology industries. For basic scientific research within the pharmaceutical or biotechnology industries, you will undoubtedly find it useful to read the same basic scientific literature that is read by academicians.

The federal government publishes a *Code of Federal Regulations* (CFR) which, as its name suggests, contains all of the federal regulations which pertain to all sorts of different industries including the pharmaceutical and biotechnology industries. This document is mentioned here simply because there are many jobs in both the private and government sectors which revolve around creating and satisfying the regulatory requirements outlined in it. However, the CFR is very tedious to read and probably will only provide a limited amount of useful job information, although it may be helpful to at least know that it exists. Many university libraries are likely to have it if you are interested.

Appendix A also contain listings of professional societies which are likely to be of interest to people in the pharmaceutical and biotechnology industries. Of course, if you are planning on pursuing basic scientific research in a pharmaceutical or biotechnology company, you will probably also want to join a professional society specific to your particular scientific discipline.

There are several other ways to identify specific companies in the pharmaceutical and biotechnology industries. The *Physicians Desk Reference* (PDR) contains information, by company, about the pharmaceutical products that are made by these companies. Obviously, the PDR contains information about drugs that are already marketed. Nevertheless, it can be a useful resource for finding information about well-established companies. Virtually all physicians, many basic scientists, and many libraries have a copy of the PDR, making it relatively easy to locate a copy. Your library may also have directories which list companies by geographic location, or by other means.

The North Carolina Biotechnology Center, the Biotechnology Industry Organization (BIO) and the Mary Ann Liebert publishing company publish directories of biotechnology companies. Their addresses are listed in the Appendix. Numerous other organizations publish directories of biotechnology as well, so you may wish to check with your university health sciences library to see what types of directories they have for biotechnology companies.

As you may have learned from previous chapters of this book, one way to break into the pharmaceutical or biotechnology industry may be to work for a contract research organization (CRO). CROs can be identified in many ways. For example, the Drug Information Association produces a *Pharmaceutical Contract Support Organizations* book which contains excellent information about contract research organizations, and some of these organizations belong to the Biotechnology Industry Organization (BIO). The Mary Ann Liebert publishing company publishes a directory which contains, among other things, a directory of contract research organizations. Contract research organizations are also likely to advertise their services in publications of interest to the pharmaceutical and biotechnology industries, such as *Pharmaceutical Executive* and *Applied Clinical Trials Journal*. Of course, these organizations can also be located on the Internet, the use of which is covered in Chapter 12.

One other way to obtain information on publicly owned companies is to obtain a copy of their annual reports to shareholders. These annual reports often contain a fair amount of technical information about the company and what it is doing. Copies of annual reports can be obtained from many business libraries, and brokers often have this information for potential investors. You may be able to get a copy of the annual report directly from the company in which you are interested.

Jobs within Academia

Academic positions are probably the easiest of all for most of us to find out about. The reason for this is because virtually all of us have at least some personal contacts in academia - we've all been there!

As is the case with other types of jobs, personal contacts are probably the best and most efficient way to learn about opportunities within academia. Of course, large universities with a large and diverse number of programs (including graduate and medical school programs) are likely to have the biggest diversity of career options.

The phone book of an institution can tell you a lot about the different offices and careers which are available in that institution. If you are unsure of what an office does, you can always give them a call! Often if you show interest, someone will be willing to talk with you about the office and the work that they do. Also, the scientific literature can tell you a lot about the people and research at a particular academic institution.

If you are looking to identify particular schools, there are several sources of information. Postsecondary schools can be identified by browsing through the *Peterson's Guides* such as the *Peterson's Guides to Programs in Biology or Medicine*, which are usually available at libraries and bookstores. Your local public or school library may have other sources of information as well.

If you are looking to use job advertisements as a source of information, academic positions at many schools are advertised in the *Chronicle of Higher Education*. Additionally, academic positions (including those in non-research, non-teaching areas such as technology transfer) are often advertised in general science periodicals such as *The Scientist* and *Science*.

Jobs within U.S. Government

Learning about jobs within the U.S. federal government can be accomplished in the same manner as in the private sector or in academia. The reason for this is because the government often administers grants to, regulates, and/or otherwise works with academia, industry, and/or other organizations. Thus, an awareness of the issues facing these types of organizations will also clue you in as to the types of jobs available within the federal government.

Identifying specific job opportunities in the federal government, just like with other employers, can be accomplished by talking with your personal contacts. Sometimes the federal government advertises its jobs in trade journals or with professional societies as well. The structure of the federal government, as well as the missions of its different organizations, agencies, and divisions, is outlined in the *U.S. Government Manual*, which can often be found at your library or can be obtained by contacting the address or phone number provided in the Appendix. Changes in government are also detailed in the *Federal Register*, a daily publication of the U.S. government which can be found in hardcopy in many libraries or which can be accessed through the Internet (**http://www.access.gop.gov/su_docs/**). Although this document is tedious to read, you can perform a keyword search over the Internet to zoom in on those topics which are of interest to you.

The federal government also provides another means for identifying specific opportunities within its various agencies and divisions. Information can be accessed by phone, mail, or electronically. You may obtain the best results by using a combination of these approaches, since each separate approach has its own limitations in the database and how it can be searched.

By phone: The Office of Personnel Management has an automated telephone service (phone number 912-757-3000) which operates 24 hours a day, 7 days a week. Additionally, your local phone book may list a local or regional number for the Office of Personnel Management. This service allows you to search for positions by title, series, or broad job category (e.g. physical and life sciences). This service also provides information on application procedures for federal jobs.

If you know which government agency or organization(s) you wish to apply to, you can call that government organization directly to obtain a listing of position announcements within that agency. The *United States Government Manual* provides information on whom to contact for employment opportunities within the different organizations of the federal government.

By mail: You can obtain a listing of openings within each government agency, office, or division by writing to the Personnel Office of the government organization in which you are interested. The *United States Government Manual* provides addresses of whom to contact for employment opportunities within the different organizations of the federal government.

Electronically: The Office of Personnel Management maintains an electronic bulletin board of position openings. It can be reached by dialing 912-757-3100 with your computer modem. Alternatively, if you have access to Telnet, you can Telnet to FJOB.MAIL.OPM.GOV from your own online provider. The electronic bulletin board allows you to search the database by a number of different methods, including geographic location and agency. This method of searching, perhaps in combination with the *United States Government Manual*, is reasonably informative and easy to understand. You can also use the File Transfer Protocol at FTB.FJOB.MAIL.OPM.GOV to obtain information about job vacancies. You can obtain federal job announcement files via Internet e-mail by directing an inquiry to INFO@FJOB.MAIL.OPM.GOV.

If you have access to the world wide web, many of these government organizations have home pages which provide information about the organization, its agencies, and employment opportunities. It is important to note that not all positions listed with the Office of Personnel Management may be listed on the home page for a particular organization. Thus, you may still wish to examine the listings of the Office of Personnel Management, even if you have visited a particular organization's home page.

The home page of the Department of Health and Human Services (**http://www.os.dhhs.gov**) provides a direct link from its home page to the Office of Personnel Management's electronic bulletin board (under "Employment"). This category of the Department of Health and Human Services home page is particularly helpful for job seekers in areas outside of the government as well.

Other: Several locations throughout the nation have computer touch screens with information about federal job opportunities. You can obtain a list of these locations by contacting the Office of Personnel Management, or the Personnel Office of any federal government organization. The Office of Personnel Management publishes a list of these locations, and of regional phone numbers for the Office of Personnel automated telephone job search system. You can also visit your local state employment service office, which also has information on federal job opportunities. Of course, if you have the time, money, and inclination, you can also directly visit the government organization in which you are interested.

Jobs within State and Local Governments

Personal contacts are again an excellent way to identify job opportunities within the state and local governments. As is the case with the federal government, an awareness of the issues facing the local businesses and other institutions in your community can help you to figure out the types of jobs which may be available within your city, county, and state governments. Simply reading your local newspaper may help to increase your awareness of these issues, as well as to provide possible names of contacts for positions within your state or local governments. Talking to people in local businesses and/or academic institutions will probably also help you to gain a better understanding of job opportunities in your state and local governments, since these organizations often must interface with the government and comply with its rules and regulations. For instance, businesses in the food industry will be able to tell you about local regulations regarding food quality and inspection (which must be made and enforced by workers within the local government). Likewise, businesses can tell you about local environmental regulations that affect them. Local law enforcement agencies may be able to help you to identify local or statewide opportunities in forensic biology. Thus, you may find that it pays, quite literally, to talk around and to be aware of the different scientific, environmental, and health-related concerns in your community.

The local telephone book or a state- or locally-run library may be a great resource for identifying job opportunities within your state and local governments. Your state and local governments are likely to have offices which can be found in the phone book and which you can visit in order to browse through job postings and apply for positions.

Many state and local governments often have at least some of the same types of positions as those which are available within the federal government. Of course, the exact types of positions available in these governments depends on the laws and concerns of your particular geographic area. Nonetheless, knowledge of the opportunities in the federal government can often give you a starting ground for beginning to explore the opportunities available in state and local governments as well.

Jobs within Other Organizations

As you have probably gathered from the other chapters in this book, there are lots of job opportunities in many different types of organizations. Because there are far too many to mention here, you are referred to the Appendix for a listing of professional societies and publications which may be useful to you. Nevertheless, to give you an example of the approaches that you can use to find information about these companies, a few sources are given here. General publications like *The Scientist* and *Genetic Engineering News* will also give you an idea of the diversity of different types of scientific organizations which exist.

The Mary Ann Liebert *Guide to Biotechnology Companies* is a great resource for identifying organizations such as law firms specializing in biotechnology, contract research organizations, manufacturing organizations, technology transfer centers, venture capital firms, and other independent organizations such as biotechnology recruiters. Additionally, some of the professional organizations listed in the back of this book produce directories of their members, or of companies which are likely to be of interest to their members. If you're looking to work for a vendor, or for companies who make certain types of products, you may wish to refer to *BioSupplyNet's Sourcebook*, or a number of other publications which list different vendors and/or supplies. Again, awareness of issues in areas of interest to you will also help you to identify job opportunities and potential employers in those areas.

One way to find out about service organizations is to look in places where they are most likely to advertise their services. For example, a company that does contract manufacturing for the pharmaceutical and biotechnology industries is likely to advertise its services in publications which are read by members of those industries. Many independent producers of laboratory supplies advertise in periodicals which are read by many basic scientists, such as *The Scientist* and *Genetic Engineering News*. Again, job advertisements can be a source of information about organizations which are potentially interesting to you. Temporary employment agencies and search firms can often be found through the classified advertisements of some newspapers, as well as in trade publications such as the newsletters produced by professional societies.

Of course, you can always ask a professor, personal contact, librarian, or professional society for other resources and directories of organizations which may be of interest to you!

Chapter 11
How to Get Hired into the Job of Your Dreams

You've found a career you desire,
But now you have to get hired.
You'll need to convince
Some hiring prince
To pay you until you're retired!

Hopefully by now you have identified at least some career paths which seem interesting to you. Now how do you now go about seizing these opportunities and make them translate into actual job offers?

If you are lucky, the opportunity of a lifetime just lands in your lap, and you never have to launch a formal job search. I was hoping that would happen to me. Unfortunately, it didn't. It didn't happen to any of my friends, either.

Thus began the dreaded job search. My friends and I loaded up on books, talked to people, and shared our experiences. With all of this information, we came up with the following techniques for conducting a job search. If you do these techniques (but not necessarily in this order), you may be on your way to a career that you want!

Please note that this chapter is written assuming that you have already chosen a career path or know what you want to do. This is a very important part of your job search. If you have not done this, then you may wish to re-read Chapters 2-8. A focused job search is more likely to lead to success.

Successful Job Searching: Selling Your Skills

All of the techniques outlined in this chapter revolve around one central principle: Selling your skills in the most optimal way. You are selling a product: *you*. Like selling any other product, you will get the best results if you highlight all of the features of the product which are likely to be of interest to the consumer.

Selling your skills in the job market, just like selling any product, involves five basic steps (Figure 17).

Figure 17

The Five Steps to a Successful Job Search

1. Assess your skills

2. Prepare your advertising and marketing tools

3. Implement a marketing plan

4. Do an effective sales pitch

5. Negotiate the price

1. Assess your skills. This means *all* of them. You probably have a lot of skills that you don't think of, or that you just take for granted. Don't do this! Think long and hard about what skills you do possess, so that you can *advertise* them. If you don't do this, your skills may go completely unnoticed by someone else. You may think they're obvious, but often, they're not!

The book To Boldly Go by Peter Fiske contains an excellent section about how to assess and sell the skills you have as a scientist. Please see the Appendix for information on how to obtain this book.

2. Prepare your advertising and marketing tools. This includes your curriculum vitae (CV) or resume and your cover letter, as well as any other materials you think may help you to sell your abilities. For instance, if you're applying for writing positions, you may wish to have a sample of your writing handy. Use any tools you think could help you to make the sale - innovative marketing tools can help you to stand out from the rest.

3. Implement a marketing plan. There are many ways to do this:
• Schmooze. Talk around and network. Make contacts.
• Apply directly to organizations of interest to you, regardless of whether you have any contacts in that organization or of whether they are advertising a position.
• Answer job advertisements in newspapers or trade journals, from the Internet, or from anywhere else you see them.
• Use headhunters or search firms.
• Use any other technique(s) you can think of. These techniques are not mutually exclusive - use as many as you can.

4. Do an effective sales pitch.

•   Try to get in for an interview. When applying for jobs, your goal is to get to *talk* to someone, particularly someone who has the authority and influence to hire you. Only then can you truly assess the employer's needs and describe to them how you can help them. If you have personal contacts, you may be able to do some of the talking *before* you even go for an interview, and sometimes, before you even submit any written materials. Use these opportunities to figure out how best to convince the employer that you would be a great asset to their organization!

•   Follow up after applying for jobs and interviewing. Send follow-up letters and/or phone calls. After interviewing, be polite and send a thank-you note. Be persistent - persistence wins job offers!

5. Negotiate the price. This is probably done most effectively once you've made the sale. You don't want to stop the employer from looking because your sticker price is too high! Nonetheless, once you've got the employer interested in you, why not try to get as high a price as you can for your talents?

Now let's cover each of these topics in a little more detail.

<u>1. Assess your skills.</u>
        You are a scientist. Inherently this means that you have a lot of skills.
        For example, you've learned to think and memorize, and to solve analytical problems. You've probably also learned to tinker with and fix lab equipment. You've learned how to do 4 hours worth of procedures in 3 hours of time.
        In fact, many people who have worked in a research lab possess virtually all the skills necessary to run a successful business. For instance, you may have had to come up with the initial ideas for the product (in this case, the product is research data). You may have had to help to obtain the "venture capital" (grants), "manufacture" the product (do the research), and "sell and market" the product (publish the work, apply for grant renewals, give talks, and generally get other people interested in the work). Throughout all of this, you may have also had a part in making sure that the lab stays "on budget" (you can't overspend grant money), managing the day-to-day operations in the lab, and having the foresight to change directions when you see that a particular research project is not going anywhere. Even if you've only participated in some of the activities of the lab, you have still gained valuable experience which is important for success in the business world.
        Think about all the computer skills, technical skills, and presentation skills that you have acquired throughout your schooling and your work experience. Then *list them*! If you have any other type of special knowledge,

even if you don't ever think you'll use it again, *list that, too*. Put it in your resume or CV. You never know what an employer will be looking for.

Getting Additional Work Experience
            Sometimes you will find that you lack certain specific skills that employers are looking for. This can be very frustrating, especially to highly educated people who have a lot of proven ability in a highly technical field. Although you may have found the cure for cancer and won the Nobel prize, you may still not get hired into a certain type of position if you don't have knowledge of some specific computer program, or some other very specific skill the employer is looking for!
            It's hard to believe. You'd think that employers would look for a person of good quality overall, rather than a specific set of skills. Often, though, this is not the case.
            The reason? Many employers are very short-sighted. They do not want to be bothered training anyone in anything, no matter how simple. Often, they wait until the last minute to hire someone, and by then, some situation has reached crisis proportions. Thus, they need you to be able to start working right away, with a minimum of training.
            This concept has two important implications:

•   Accumulate (and emphasize) as many skills as you can. Actual work experience can help you to acquire more skills. You can also learn them on your own if you have to (anyone can teach himself WordPerfect).

•   You will probably never have exactly the credentials that your employer is looking for. That's because employers often have unrealistic expectations. So don't be afraid to apply for jobs for which you do not have all of the credentials. It's rare for an employer to find a person who does! Emphasize the credentials that you do have, and how they could relate to the credentials that the employer is seeking. Offer to learn on your own time.

            Internships, college work-study, co-ops, summer jobs, and other work experience (including volunteer work) are excellent ways to acquire some of the specific skills that a potential employer may be looking for. These opportunities are often available in government agencies, the private sector, and in academia. So be sure to ask about them.
            Okay, so now you realize that you do have a lot of skills, and you've assessed what they are. Now what do you do to advertise them? The next sections cover the tools you will need for communicating these skills to an employer in a meaningful way.

2. Prepare Your Marketing Tools
            The purpose of your marketing tools is to emphasize those skills which are relevant to the position for which you're applying. The way you present your

skills is important to your job search. You'll want to present them in a way that makes it obvious to an employer how your skills can help the organization. This applies to your interviews as well as to your written marketing tools.

To illustrate my point, let's suppose you are lazing around the beach. Suddenly, you spot a drowning swimmer. Wanting to help, you swim out to him.

You are now trying to convince him to let you help.

"I want to help you", you say.

"You're not a lifeguard!", the swimmer screams.

"I am a strong, experienced swimmer. I am also a certified emergency medical technician. I can help you," you respond.

The swimmer may still not cooperate with you because he's panicked and hysterical. He may struggle with you and fight you because he doesn't trust you. At least, however, *you've* tried your best to win his confidence. Even though you may not be a lifeguard *per se*, you've highlighted those skills which enable you to get the job done.

This same strategy often works with employers. After all, employers hire people in order to get them out of predicaments. They are like drowning swimmers. They are desperate for help with a problem which needed to be fixed yesterday, or a project that needs to go out the door tomorrow. So ask what it is they need (if you don't know already), and tell them how you can help them. Don't make them figure it out for themselves - they may never do so.

You will still want to include *all* your skills, because you never know what an employer is looking for. But try in particular to emphasize those which you regard as being the most important for the job you are applying for. This same guideline will pertain to your interview. Interviewing is discussed later in this chapter.

Since there are so many books written about resume and cover letter preparation, this book will not go into the gory details. Instead, you are urged to refer to one of the books listed in the Appendix, or some other career book of your choice, for the details of how to prepare and format a resume. The book To Boldly Go by Peter Fiske is especially recommended, since this book was written specifically for scientists. Information on this book is listed in the Appendix.

Based on what I said earlier, though, and based on these books, there are a few points I would like to make:

•   There is actually a lot of room for flexibility and creativity in preparing your resume or CV. That is why different books often suggest different formats. Remember that the basic point is to sell your skills in a relevant way. So just pick a format that you think will work for your level of experience and the type of job you're applying to.

•   Make sure that your resume is understandable - don't use all sorts of techno-speak unless you're sure your audience will understand it! Otherwise, try to figure out a way to explain how your technical skills can benefit your employer in a simplistic way.

•   Use your cover letter to further highlight information about yourself that is likely to be of interest to the particular employer to which you are applying. Supplement your resume and cover letter with other materials that you think might convince an employer to hire you. Be creative!

•   A note on electronic resumes: You are probably aware that it is now possible to submit your resume over the Internet. This is fine; however, just keep in mind that your resume must still be formatted and neat in appearance. This can be tricky - you will want to ensure that the formatting does not get altered by its transmission over the Internet. If in doubt, you may wish to check with someone who can help you with this.   You may also wish to consider the security of information over the Internet. If that's important to you, you may wish to send your resume by some other means.

•   Don't become paralyzed by the quest for the "perfect" resume, cover letter, or other job search material. There is no such thing! It's more important that you get your materials into circulation than it is for you to keep trying to perfect your materials. No one will know you're looking for a job if you never send anything out!

      If you do feel paralyzed by the fear of a mistake or of having less-than-perfect materials, perhaps the following story may be of some comfort to you. A friend of mine was convinced that there was a "perfect" resume and cover letter. She went all over the place asking people's opinions and reading career books. Based on their varied advice, she changed her resume more often than Madonna changed hair colors. In the meantime, because of the time she was spending on preparing her job search materials, she was not applying for any positions.

      She finally decided to redirect her efforts. She put more effort into identifying potential job leads. She then followed up and submitted her materials to them, no matter how "imperfect" her materials might be. Although she tried to be careful, she was scared to death that someone might find a mistake. She pressed on anyway.

      Lo and behold, early on in her job search, my friend made a mistake. She found that she had misspelled the last name of the department head in a cover letter addressed to him. She had read all the Dale Carnegie teachings which emphasized the importance people place on their own names. She believed that she had made a horrible mistake.

      Nevertheless, she mustered the nerve to call the department head. "I would like to discuss the Rocket Scientist position available in your company. Recently I sent you a letter with my materials. I'm sorry for misspelling your name on that letter. By the time I realized my mistake, it was too late."

      "Oh, that's okay," the department head replied. "Let's talk about the position. What do you think you have to offer us?"

      Now that the conversation was on this track, the applicant was able to talk about her qualifications and potential contributions to the company. In the end, she was awarded the job.

The moral: Mistakes are correctable. Of course, you don't want to deliberately do sloppy work. But don't let the fear of making a mistake paralyze you from applying for jobs.

There is another moral to this story. That is, the approach to identifying and following up with job opportunities is actually more important than the exact appearance of your resume and cover letter. A fancy but misdirected cover letter and resume is unlikely to produce results. However, an average resume, accompanied by superb follow-up, can win you job offers.

3. Implement a Marketing Plan.

The marketing plan is probably *the* most important determinant of whether you get a decent job. Sure, you have to do well at your interview, and your written materials must be good, but none of that stuff will do you any good without a good marketing plan!

For instance, suppose that during your graduate work, you performed a set of experiments that suggest a cure for a disease. You strongly suspect that your work has commercial and humanitarian value. As your experiments progress, you are more and more convinced of the work's value.

However, let's say you don't publicize the work. Maybe you're lazy. Maybe you hate to write. Maybe you hate to give talks. So meanwhile the data aren't doing anyone any good.

How many Nobel Prizes do you think you'll win with this approach?

Your job search operates on the same principles. Your talents won't get you a job if you don't communicate them to an audience who will appreciate them. If you want to get a decent job, you need to let people know how you and your work can help them!

So let's talk about how to market your skills. First, you'll have to identify your target audience. This involves identifying prospective employers and/or positions in which you'd like to work. This topic is the focus of many of the other chapters in this book. The book What Color Is Your Parachute by Richard Nelson Bolles also contains some excellent generic hints on how to identify job opportunities and people who can help you.

Nevertheless, here are some techniques that you might find helpful for identifying *specific* target audiences to whom you can market your skills:

Networking

Why is networking so popular? It is because networking is a time-honored technique that actually works. And if it works, you work! So why is networking such an effective way to find a job?

• Networking greatly multiplies the number of eyes and ears that you have on the lookout for you! This will lead to a greater number of opportunities, just by statistical chance.

•   Networking can put you in touch directly with the person for whom you could be working. This gives you the opportunity to sell your abilities to the person who is most likely to appreciate them. It also can make you stand out - if you already have a contact, it can inspire the "right" person to pull your application out of the pile and give it some extra attention.

•   When you use personal contacts to help you to find a job, you come complete with references already! You (or your connections) are not a complete stranger to the person who is doing the hiring. This is comforting to people.

Do not underestimate the number of techniques you can use to network. The more techniques you use, the better - that's more eyes and ears looking out for you.   Talk to your friends, relatives, teachers, and trusted co-workers or business contacts. Join clubs and professional societies, and get to know people. Even your present boss may be able to help (certainly, though, you will want to use your judgment on this one). Sometimes it is a well-known fact that you will eventually be looking for a job. If you are working in one of these arrangements, make use of the contacts that you establish there.

Here's a technique that I and many of my friends have used with success. We went through the membership directory of relevant professional societies, seeking names of people who worked in job titles (and in companies) we thought we'd be interested in. We then got on the phone and called these people. It was hard, particularly for those of us who were shy. But it worked.

If you decide to make "cold calls", it is probably easier to start by calling some of the more junior people first. Although they may not have as much power as a senior-level person to hire you, they offer several distinct advantages. Firstly, they are often more likely to be at their desk. Secondly, they are often less intimidating. Thirdly, they are also more likely to be sympathetic with your situation in trying to find a job. After all, it hasn't been that long since *they* themselves had been looking for a job.

Other ways in which to identify people whom you can contact is through an organization's annual report or from the scientific literature. Just keep in mind that it is always better to have a personal contact to write to or call, rather than to send in a job application to no one in particular. Make every effort to identify a name, even if it means just asking for the head of a department and finding out his or her name that way. You can probably think of plenty of devious schemes for doing this. You may even have some fun at it.

Getting more work experience is also a very effective form of networking *and* identifying other good job opportunities!

Applying Directly to Companies of Interest to You

Sometimes you don't have any contacts in organizations which interest you. Even if you do, you may want to supplement your networking with some other approaches.

Here's a technique that may work for you. Identify some specific organizations for which you may want to work, then go there and apply directly to them for a job. The main feature of this is that **you do not have to wait for an organization to announce a vacancy before you apply for a job there**. Actually, you may be better off if a position is not presently being advertised. That is, if the organization has not announced a vacancy, they are not receiving a thousand resumes for a position. This may put you in the forefront of the organization's mind for a suitable position when one arises. Perhaps you can even talk the organization into creating a position for you!

Again, you can try to identify a contact there. Maybe someone in the organization belongs to a professional society for which you can obtain a directory. Even if you do identify a contact there, you don't have to wait for this contact to refer you - you can go ahead and apply anyway!

Answering Job Advertisements

Answering job advertisements is probably the most popular way for people to look for a job. And indeed, it works for many people.

Some people, such as Richard Nelson Bolles in <u>What Color Is Your Parachute</u>, believe that answering job advertisements is less effective for landing a job than some of the other techniques described in that book (including networking, which he states is the best technique). There definitely is merit to his opinion. In fact, looking through job ads can take time that could be spent in other ways. It also can be harder to compete for the jobs because so many applicants may apply for each position in response to a job ad. The first two techniques listed above avoid this situation.

Nonetheless, if you do answer job ads, you may have the advantage that the company is actually looking for someone at the time you apply. And they may want to bring in the person quickly! Although Bolles points out in his book that some organizations run ads just to assess the job market and the quality of applicants, many organizations really are looking for someone to fill a position when they run an ad.

More information on where to locate job ads is found in Chapter 10.

Using Recruiters, Headhunters, Employment Agencies, and Other Search Firms

Recruiters, headhunters, employment agencies, and search firms are inherently a wonderful concept for a job seeker - someone else finds you a job.

Let's read more closely between the lines. You may think that these firms have the connections and contacts to be able to place you into a terrific job. And in fact, they might. However, you should be aware that they are *not* career counselors. They are in the business of placing you, not making you happy! Your happiness is still *your* responsibility, and you may still have to do a fair amount of research to be able to guess whether you'll be happy in a certain position versus whether you'd be better off with something else.

This topic, as well as the different types of organizations which exist, is covered quite thoroughly in <u>What Color Is Your Parachute</u>. Since it is beyond the scope of this book to cover all of the different aspects of search firms, headhunters, and employment agencies, you are urged to refer to that book for a better understanding of how these firms work.

If you do choose to use a headhunter, search firm, or employment agency, you will soon discover that different firms specialize in different markets. For example, some firm may specialize in highly technical, bench-level positions, whereas another may specialize in clinical research. Some larger firms may cover a number of different markets. Some may place you into temporary or contract positions, whereas others may deal with more permanent positions. Some firms place "entry-level" people, and many do not.

By now, you're probably getting my point. You may wish to ask each firm what their specialty areas are and what types of positions they place people into. Further, you may wish to identify firms in your field of interest by checking out resources in the particular specialty area(s) in which you are interested. Many times, trade publications such as *Genetic Engineering News* or *The Monitor* (published by the Associates of Clinical Research Professionals), or publications of the professional societies themselves, will have advertisements for search firms and employment agencies. These firms also often advertise at meetings of professional societies.

If you are looking for a job in a particular geographic area, you may wish to check out that area's local newspaper for ads of search firms which place people in that area. The local phone book can also be a source of information for identifying local search firms. Please note, however, that you do not necessarily have to use a local firm in order to identify job opportunities within a given geographical area. Many firms conduct searches on the national and even the international level. Thus, a firm in California may well be able to place you into a position in Virginia.

Sometimes using a headhunter or employment agency is all you need to get a break into the world of work. Many people have accepted temporary jobs through such firms, which gave them the experience and contacts to land a permanent position. Since a good number of employers will hire people only through these firms for temporary or contract positions, the job seeker would not have landed these positions on his or her own. Thus, headhunters and employment agencies can definitely serve a purpose for you, if you're careful about how you use them!

If you do decide to use a search firm or employment agency, you may wish to be aware of one more thing. It is helpful to tell the firm exactly what you want (at least as much as you can), and to sell your skills to the firm just as you would to an employer. That is, just as an employer would want to know how you would help them in a particular position, the firm needs to know that, too. Otherwise how can it recommend you for a certain type of position?

4. Do an Effective Sales Pitch

The best way to get an opportunity to sell your skills is to get your foot in the door for an interview. Thus, all of your efforts should be directed at this goal. Once you get in for an interview, you have the opportunity to interact with a person who has the power and authority to hire you, and you have the chance to find out exactly what the employer is looking for. Only then can you truly sell your skills in a way that the employer will find irresistible!

A Note on Following Up after Applying for a Job

Often when you apply for a job, you don't hear from the employer for weeks and weeks. You sit there and stew, thinking that you'd get as quick a response by FedExing your materials to Mars. In fact, you may be right - you get no response at all.

So what do you do then?

Maybe you try to protect your feelings by trying to pretend that your resume got lost in the mail. Or maybe you keep hoping that the employer will call you soon. Worse yet, perhaps you sink into the depths of despair and quit applying for positions, figuring that you'll never get a job.

Don't do this! Instead, be proactive and follow up. Maybe you can send another follow-up letter, or maybe you can call the employer.

For those job ads that say "no calls, please", you can still continue to express interest in the position. You can visit the employer in person. If you do this, you can try to get someone to chat with you about the job, its responsibilities, and what type of person and skills the employer is looking for. This information will help you to sell yourself.

In any event, you're probably getting the picture by now. You'll need a lot of persistence throughout your job search. Don't take this personally. Most employers are busy, and even though they want to hire someone, it's low on their priority list. Interviewing and hiring takes time, and many employers are struggling daily just to get through the things they need to accomplish. After all, a heavy workload may be why your potential employer is looking to hire someone in the first place!

Inevitably, there will probably be times when you will worry about annoying your potential employer. Of course, this can happen. However, more often than not, the employer understands the situation and actually respects your persistence. Persistence in your job search usually means persistence and diligence in your work, and most employers want that!

The Interview

The purpose of the interview is to convince an employer beyond a doubt that they want to hire you. Thus, your job is to sell your talents and abilities to the employer in a way which they will find irresistible.

Remember that the employer is like a drowning swimmer. Your goal is to figure out how to convince him to let you help him. So be sure to highlight those skills that are likely to be of interest to him and relevant to his current situation.

So how do you know which skills are most relevant? Just remember that an interview is a two-way street and an exchange of ideas, and use this to your advantage. Take your cues from the employer. In fact, encourage the employer to do a lot of the talking - and *listen*! This'll be the only way you will truly find out the nuances of the position, as well as what the employer *really* wants. Besides, people like to talk and will think you're a great conversationalist if all you've done in fact is to listen.

Again, interviewing skills are the subject of many books. Thus, I will not repeat all of the same information here. Instead, I urge you to refer to one of them. Nonetheless, there is one universal word of advice that can be gleaned from all of these books - be as prepared as possible! Come prepared with answers to questions you think you might be asked, as well as some questions of your own. Your career counselor at school or one of these books may be able to help you to identify some questions that many employers ask during an interview.

Finally, here's a word of encouragement. Sometimes you may think that an interview is not going well. But before you get too nervous, let me relate to you a story about what happened to one of my friends during a job interview. The interviewer kept pressuring her and making her uncomfortable. Finally, my friend decided that she had had enough. She turned to her interviewer and said, "If your company treats its employees like you are treating me in this interview, I do not want to work for your company." It turns out that the interviewer was deliberately making the interview unpleasant to see how she handled pressure (some organizations do this and call it a "pressure interview"). Anyway, the interviewer was actually impressed by her ability to speak up and not be pushed around. She got the job!

Following Up After the Interview

Everyone likes polite people. This alone can win job offers. People like to work with others who respect them, who are polite, and who are easy to get along with. So why not give the employer yet another reason to hire you?

There is one other selfish reason for writing a thank-you note. It gives you one last opportunity to sell your talents. Perhaps you didn't get a chance to say certain things or didn't think to say them during your interview. So state them now! Tell the employer how happy you were to have the opportunity to meet them, and tell them how and why you're convinced you'd be good for the organization.

Unless explicitly told otherwise, you may also wish to state when and how you'll follow up with the employer again. Doing this shows interest and initiative - so don't leave the ball in his court! Then do what you state in your letter. Call or send another letter or an e-mail. Keep following up either until you get the position or until you're absolutely convinced that you are definitely not a candidate for the position. Remember to be polite, no matter what the outcome.

Lastly, get that thank-you note into the mail as quickly as possible after the interview! That way, you're fresh in the employer's mind. Also, if by some miracle the employer does actually make a decision quickly, they will already have received your thank-you note by the time they make the decision. This can only help you!

5. Negotiate the Price

Let's not forget that the primary reason that many of us work is to make money! Even if you think that the money is not all that important, you may wish to reconsider. In fact, your overall compensation package does matter.

A competitive salary does more than ease your financial burdens, thereby making it easier to go in to work every morning feeling positive about your work and able to concentrate on your job. It also affects others' perceptions of you. Did you know that studies have shown that consumers often buy a more expensive product (rather than a cheaper product of the same quality) simply because they perceive that a higher price is indicative of a higher quality?

This same psychology is at work with your salary. The more you get paid, the more value you are perceived as having in the marketplace (both financially and otherwise).

So how do you go about negotiating a competitive compensation package? Again, there have been many books written on this topic, so I will not go into much detail here. However, here are some pointers which you may find helpful:

• Do not talk salary until you think that you have made the sale (i.e., until you're reasonably sure, or better yet, absolutely sure that the employer wants to hire you). You probably don't want to lose a job offer just because you seemed more interested in the money than in the job, or because you naively gave too high a number. If you are under considerable pressure to give a number, try to give a range rather than an absolute number, and say that the acceptable figure depends on a number of things (for example, promotion opportunities, flex time, vacation time, sick time, etc.).

• Know your approximate worth in the marketplace before going in to a job interview. That way, you are prepared to talk about salary if the topic comes up and you cannot avoid it. There are many ways of finding out this information - you can call up people with similar job titles (for instance, you can get the names from a professional society directory), introduce yourself, tell them about your situation, and ask them approximately what you could expect in terms of salary for a certain type of position. You could also call Human Resources Departments at several organizations similar to the one you are applying to, and see if they will give you the pay scales (often, they do have this type of information). This information really can pay off - a friend of mine knew exactly what she could be making (almost down to the penny) in a similar industry to one she was applying to. When the talk of salary came up, she gave a range. When her potential employer said that she would be making closer to the bottom end of the range, she acted surprised that it wouldn't be more. She mentioned exactly where she was getting her information from. She said that she had been told that if she were to start at company X, she would be making Y amount of money. By doing this, she probably got $5000-$6000 more than the company was originally planning on offering her. That's not a bad payoff for a little bit of time and effort she spent to educate herself about the job market to which she was applying!

•   Keep in mind that the time for salary negotiation is *before* accepting the position. Once you get into a particular job, your raises will most likely be based on your present salary. It will then be more difficult (without changing organizations) to obtain a significant leap in salary. Thus, the present negotiation can determine your salary for years to come!

•   Remember that your overall compensation package includes more than just your salary. If the organization cannot pay you more, perhaps you can negotiate more vacation time or flex time, or work fewer hours per week.

•   Finally, for all this talk about salary negotiation, it will be a moot point with some organizations (particularly large companies and government agencies). Some of these organizations will have strict formulas, based on your experience and education, which they use to decide what you will get paid. There really is no room to negotiate in these types of organizations, so sometimes you have to accept that.

Chapter 12
Your Virtual Job Search: Using the Internet

When searching for more information
Or looking for jobs by location,
A click of the mouse
Right from your own house
Can lead you to your destination.

The Internet has all sorts of wonderful information which can help you with your job search. In "virtual reality", you can go all sorts of places, free of charge, without ever having to start your car. Thus, if you're a student or a worker, or if the word "broke" describes either your financial situation or your car, this chapter's for you! Just about every student and worker has some form of access to the Internet, either at home, through your employer, or at your university or public library.

In writing this chapter, I once again realized how much technology has affected our methods of job searching. For years, older generations have pointed out how easy things are "nowadays" compared to when they were "our age".

<u>Two Generations Ago:</u>   "When I was a kid, we had to walk, 25 miles through the rain, sleet or snow, to get to work or school. It was uphill both ways. There were no such things as cars or buses back then, and telephones weren't part of every household. Heck, we didn't even have shoes in those days. People either worked on the family farm or traipsed all over town to find a job. We came home physically exhausted at the end of every day."

<u>One Generation Ago:</u>   "Kids these days have so many advantages. When we were growing up, we had to drive the old, beat-up family car to get to our jobs. We had to make sure to be home on time so that others had access to the car.

Nowadays, not only do kids have their own cars, but they have computers as well. They don't even have to leave the house in order to look for a job."

It is probably safe to assume that the present generation, too, will complain about how hard things were for us. I wonder what exactly we'll say. Maybe it'll go something like this:

Our Generation: "When I was looking for a job, I assumed I'd have to work for a living. There was no such thing as a 'virtual' job. Nowadays, people don't even have to show up to work in order to collect a paycheck. They can stay home, fiddle with their computers, send something over the modem every now and again, and call it 'work'."

In all reality, technology *has* changed a lot of things. Jobs have changed, as well as our methods for finding them. To this end, the Internet can be a very powerful source for gleaning information about virtually every career path. The problem is, how do you know where to go? Secondly, what do you do once you get there?

How to Use the Internet for Your Job Search

The Internet is a complicated, chaotic, constantly changing forum for the exchange of information. Luckily for us, for each nerd that invents a new tool, there's another nerd who writes about it. Since it is beyond the scope of this book to give basic instructions on the tools and uses of the Internet, you are urged to ask a friend or to refer to one of the resources mentioned in this chapter or in the Appendix for more information, if you need it. There is also help online on the world wide web at **http://www.spry.com**, which provides an overview of the different tools and applications of the Internet.

One source worthy of particular mention is "*A Biologist's Guide to Internet Resources*", by Una Smith. This site, the address of which is listed in Appendix D, not only describes the different tools of the Internet, but it also keeps track of Internet sites which are likely to be of particular interest to biologists. It describes websites, Usenet newsgroups and chat groups, and other areas of the Internet relevant to biologists. Its originator, Una Smith, should probably be initiated into biological sainthood.

Another person who fits into that category is Mary Ann Liebert, whose company publishes many things including *Genetic Engineering News* and a *Biotechnology Internet Address Book*. The American Society for Microbiology also publishes a book about the Internet, and a visit to your local computer store may also help you to gather hints on how to navigate the Internet.

So, how can the Internet help the job seeker? It actually has a number of different uses and applications for helping you to find and/or land a desirable position (Figure 18).

Figure 18

Internet Uses for the Job Seeker

- Keeping up with current events in your field
- Searching for information on specific organizations or types of organizations
- Identifying and placing job advertisements
- Applying for positions and submitting your resume online
- Making new personal contacts
- Keeping in touch with your existing personal contacts.

Now let's cover each of these in a little more detail.

Keeping Up with Current Events in Your Field

The Internet offers many tools to enable you to keep current in your field. There are online publications, websites, newsgroups, bulletin boards, chat groups, mailing lists, and the like. There are files of information that can be accessed using Internet tools such as Telnet or file transfer protocol (ftp).

Okay, you already knew that. Alternatively, if you didn't already know that, please go online and either visit the website **http://www.spry.com** or see the online resource *A Biologist's Guide to Internet Resources* by Una Smith (address given in Appendix D). Alternatively, you can refer to a book such as *The Internet for Dummies* (please see Appendix C), for a basic explanation of the Internet and its tools.

Anyway, once you are familiar with the Internet and its tools, you still need to know where to go in this chaotic mess. That's where the online guide *A Biologist's Guide to Internet Resources* by Una Smith can help you.

Appendix D also lists a number of different online sources which may interest you. These sites were chosen from listings in trade publications. For additional sites in a particular field, you may wish to look in the trade publications which are closest to your particular field of interest. Many of these publications now have listings, and even critiques, of interesting Internet resources.

Additionally, there is help available online. There now exists a group of websites called "search engines" which, as their name suggests, help you to search the Internet for sites which may be of interest to you. There are even online lists of search engines! Some of the more popular search engines, and some of the sites which list these search engines, are listed in Appendix D.

A note on search engines: Not all search engines are created equal. Databases differ, as do the searching techniques that different search engines offer. Because of these differences between search engines, you will probably want to experiment around with different engines when looking up information on any given topic.

As you probably know, the world wide web is not the only place to go and tool to use on the Internet. Resources have also evolved to help you to find information in other places. For instance, there is a site which lists all the gophers in the world, and there is a site which tries to keep track of all active Usenet newsgroups. These online resources are also listed in Appendix D. For more of an explanation about these different tools of the Internet, please see one of the references mentioned earlier in this chapter.

One very popular method for finding interesting sites on the web is by using hypertext links from other sites in related topic areas. Hypertext links are text items which you can click on to link you to another Internet address relevant to that topic area. For instance, if you are looking for contract research organizations which serve the pharmaceutical industry, you can probably find hypertext links on websites which are related to the pharmaceutical industry. This technique may take a lot of time but can also yield some interesting information!

Obviously, there are many ways for gleaning information from the Internet. You can do patent searches, look at gene databases, read articles, look at bulletin board listings and Usenet newsgroup topics, and do many other things to try to learn about general current events in your field of interest. You'll have to experiment around for the methods that best fit your needs.

Finding Information on Specific Organizations

Maybe you already have an interview with a particular organization, and you wish to find more information on that organization. Or maybe you are interested in applying to a particular organization and would like more information on it. Maybe you're looking for information on a specific professional society that you've heard about.

Many of these organizations have a presence online, in the form of a "home page" on the world wide web. Alternatively, these organizations may have a gopher system, or you may be able to access their computers via Telnet and transfer files via file transfer protocol (ftp).

There is one particularly high-tech method for locating specific organizations on the Internet. This method is aptly called the "guess method". The guess method works like this - to find the organization on the world wide web, you type in the name or initials of the organization as part of its URL (uniform resource locator) address. For instance, many companies can be found on the world wide web by pointing your web browser to http://www.companyname.com, in which the name or initials of the company are substituted for companyname. Likewise, many government organizations can be found by pointing your web browser to http://www.governmentagency.gov, in which the name or initials of the government agencies are substituted for governmentagency (more often, it is the initials - for instance, the U.S. Patent and Trademark Office would be http://www.uspto.gov, and the U.S. Department of Agriculture would be http://www.usda.gov). Professional societies can often be found by pointing your web browser to http://www.professionalsociety.org, in which the name or initials of the professional society are substituted for professionalsociety, and educational institutions can often be found by pointing your web browser to http://www.universityname.edu, in which the name or initials of the educational institution are substituted for universityname.

If you're interested in getting into the gopher systems of these organizations, the same technique often works, but with slight modification. For instance, to find an academic institution, you would type in gopher://gopher.universityname.edu, instead of http://www.universityname.edu, from your web browser. You can try the same thing with Telnet, although it may be slightly less effective with Telnet than on the world wide web.

You may also be able to use the Internet to find articles in which a particular organization is mentioned, or patents that the organization has. That is, some sites allow you to do a search of their databases in which you can use that organization's name as a keyword. This technique can be an interesting way to find out about news related to that organization, and current events pertaining to that organization.

Identifying Position Postings and Specific Job Advertisements

The Internet can be a tremendous resource for identifying job opportunities in your specific field of interest. Job ads can be found in so many ways that it's hard to describe them all! Nevertheless, here are at least a few of the popular ways.

For starters, there are some search engines which are designed specifically for job seekers. These search engines can find you all sorts of job advertisements. Often, they allow you to search their databases by subject area, geographic area, and sometimes in other ways as well. Some of these search engines are mentioned in Appendix D.

There are also a number of biology-specific resources which post job advertisements. For instance, the home pages of some professional societies will allow you to connect to their job advertisements via hypertext links. You can view the job ads published in *Science* in this way. Some organizations, such as the U.S. federal government, will also allow you to access their job database via Telnet (see Chapter 10). Universities may list jobs via their gopher systems. Some organizations may even send you lists of job advertisements via e-mail. Basically, job advertisements are all over the Internet - if you go online, it will be hard to miss them!

Use the Internet:
Get Your Results in Real Time!

Applying for Positions Online

Throughout all of your job searching, you will probably find opportunities for which you can apply online. This is a convenient, cheap, and quick way to get your CV or resume out to a number of different employers. You can also post your resume online on a public place and hope to get "hits" from prospective employers.

Just a quick word of caution (which was also mentioned in Chapter 11): When submitting your CV or resume online, you may wish to be careful that the person receiving your resume will receive it in a properly formatted form. Sometimes transmission of materials over the Internet can alter their formatting. If you do not know how to prevent this from happening, you may wish to seek the advice of someone who does. Your materials are still expected to be neat and formatted, even when submitting them online! You might also want to consider the security of transmission over the Internet, as well as issues regarding your personal privacy.

Making Personal Contacts via the Internet

In case you have not yet gotten a particular concept by now, I will reiterate it. Personal contacts are a very effective way of getting hired into a good position of your choosing.

Now you're really in luck, because the Internet can help you out with this! The Internet provides you with a cheap and convenient way of "meeting" people from all over the world! You can participate in Usenet newsgroups and chat groups. You can even write articles that you post online, hoping that people will contact you about it. You can send personal e-mail to other individuals within the newsgroups, chat groups, and/or mailing lists. You can even send e-mail to individuals whom you are interested in getting to know, for example, because of a scientific article you enjoyed that was written by that person. E-mail is a great way to get to know another person in a relatively non-threatening way.

Several organizations, such as the American Association for the Advancement of Science and the Young Scientists' Network, have chat groups and newsgroups which may be of interest to you. Their Internet addresses are listed in Appendix D.

Keeping in Touch with Acquaintances via the Internet

Last but certainly not least, Internet e-mail can be a great way to keep in touch with acquaintances you already have. You can follow up with people you've met at scientific meetings, and as you probably already know, you can keep in touch with your friends. You never know who may someday be in a position to hire you!

One word of caution: The Internet is a tremendous resource for finding information and keeping in touch with people. However, it is not the only job-searching tool available to you. In fact, it may not always be as effective as some other job-seeking tools. Just think about how easy it is for an employer to ignore an e-mail by just hitting the "file" or "delete" key. A phone call or a personal visit is often harder for an employer to ignore. Additionally, since employers are people too, many of them may still prefer to put a voice or a face with a name. Especially in these instances, a phone call or personal visit may really help your job search.

Finally, you are probably well aware of the effect of the Internet on time. It really is easy to spend a lot of time on it. Let's face it, a trip through cyberspace is like a trip through real space. The only problem is, time seems to speed up, rather than slow down, while you're on it (even if you *are* traveling at the speed of light). There is a vast amount of information on it, and it is fascinating. It is therefore easy to get sucked into black holes and to wind up light years away from your original destination. Thus, if you choose to use the Internet for your job search, it may be important for you to remain focused on what you are trying to do and to limit the amount of time you spend on it, so that you can spend time doing the other essentials of job searching (such as actually applying for positions and making phone calls). Nonetheless, the Internet can be a great resource if you know what you want and can figure out how to get it!

<u>Specific Resources</u>

Specific Internet sites which may be of interest to biologists are listed in Appendix D.

Chapter 13
From Learning to Earning: How Much Education Do You Need?

Bachelors, Masters, Ph.D.,
What degree is right for me?
With bills to pay and mouths to feed,
Just how much schooling do I need?

How much education do you need to embark on a successful and satisfying career path?

When I first went to college, I had no what it was all about. It finally dawned on me on a beautiful fall day during my senior year of college. I was enjoying a stroll across campus with a friend and discussing life after graduation.

"I'm going to graduate school," I proudly announced.

"I've applied to graduate school too," my friend replied. "But I'm also using the college's placement service to look for a job."

"A job?", I asked, stunned. "Why would you want to get a job?"

"Well," she stammered defensively, "I wanted to see what was out there. It's kind of a curiosity thing."

"What would you do if you found something you were actually interested in?", I asked.

"I would take it," was her simple reply.

I realized at that moment how horrified I was by the prospect of having to get a full-time job after college. It seemed so confining to me, so prison-like. After getting used to changing classes on a beautiful campus, I didn't want to be confined to an office, answering to a prison warden otherwise known as a "boss". I didn't want to glumly march off to work before dawn in the morning, only to return long after dark at night, only briefly catching glimpses of sunlight through the boss' window. I definitely did not want to have this kind of lifestyle in exchange for less money than I had paid for a single year of college tuition.

Did I really go to college to get a job?

When I more fully examined the situation, I realized that I had gone to school for a number of different reasons:

Reasons to Go to School

- To get an interesting and well-paying job.

- For prestige, respect, and/or a sense of accomplishment.

- Because learning and college was fun.

- To please my parents.

- So I didn't have to get a job! That way, I didn't need to figure out what I *really* wanted to do for the rest of my life!

Finally, as the end of my graduate schooling drew near, I faced the fact that I was supposed to be going to school, at least in part, as a means to an end.

I WAS GOING TO SCHOOL TO GET A DECENT JOB!

Of course, it would have been nice to have really thought about this before blindly embarking on an educational program.

Specific Types of Degrees

The following are some of the different degree programs you might consider for the types of careers mentioned in the previous chapters of this book. Some examples of positions and their educational requirements are listed in Appendix A. One thing you'll find is that some career paths have very defined educational requirements, whereas others allow considerably more latitude. Oftentimes the amount of education required for a certain type of position is not fixed in stone.

The Bachelors Degree

The bachelors degree is considered by many people to be the minimum ticket into most types of jobs. It seems to be society's standard by which people are judged to be "educated".

Since so many people in the workforce now have a bachelors degree, you may feel that you need to rush off to graduate school to "set yourself apart" from the competition for promotions and raises. You may see people with bachelors degrees who are performing jobs that were once performed by people with only a high school diploma. Thus, you may rightfully question how a bachelors degree can provide you with adequate opportunities for advancement. You worry that you'll get stuck in a low-level position for the rest of your life.

Indeed, there is no guarantee that this won't happen. But before you reflexively enroll in a graduate program, consider the case of my friend Beth. After graduating from college, Beth accepted a job as a secretary in a small biotechnology company. She figured that she would continue on for more education at some point, but for now, she wanted to get some work experience.

At this small biotechnology company, Beth became very well versed in the company's projects. As the company grew, Beth moved through a series of promotions with increasing responsibilities. She managed projects, became involved in regulatory work, and became involved with strategic planning. She soon became quite knowledgeable about the biotechnology industry in general.

Because of Beth's new level of exposure and responsibility, she no longer had to look for a job. People were calling her with employment opportunities and job offers! Beth changed jobs several times, each time increasing her level of responsibility and expanding her base of skills.

Beth never did return to school. She was busy expanding her career! Now 48 years old, she is the vice president of a small company, and she loves her job. She simply does not have time to return to school. She doesn't need to.

In fact, there are many positions in which work experience can substitute for education. If you can bring value to an organization in a particular role, the organization may not care about what specific degree or level of education you possess. In some cases, you may be able to succeed even without a bachelors degree! Just think about how you vote for the President of the United States. Do you vote for a person based on the degree he possesses, or based on what you think the person can do for the country?

Selecting the right type of undergraduate training can help your career to move in a direction you wish to go. You can take certain combinations of courses, and/or do appropriate internships, which can steer you in a certain career direction and obviate the need for further graduate education. For instance, if a career requires a special certification (such as medical technology or forensic science), you can look into the possibility of acquiring this certification as part of your undergraduate training. Perhaps if you know you will be interested in combining your biology skills with those of another discipline (e.g., computer programming or public relations), you can take additional coursework or do an internship in that particular area. You may even wish to have a double-major or a minor concentration in that subject area.

The Masters Degree in a Basic Scientific Discipline

You may get into the working world with a bachelors degree, only to find that you really wish you had more education. Perhaps you feel that your lack of higher education is causing you to lose out on certain opportunities. A masters degree may be just what you need to break this cycle of being "unqualified" for a job and to get you started on a career path that you desire.

The Masters of Science in a biological discipline can offer you a considerable number of different career opportunities that may not be available to people with bachelors degrees. The masters degree is a chance to gain some expertise in a specific scientific area (such as microbiology, biochemistry, pharmacology, toxicology, or epidemiology), without being quite as pigeonholed into a specific research program as someone with a Ph.D.

The M.S. degree is often obtained "on the way" to getting a Ph.D. degree. Because of this, the M.S. in a biological discipline is often viewed as a consolation prize for people who were unable to get their Ph.D. degree. This viewpoint is unfortunate, because the masters degree definitely has some advantages over the Ph.D. degree. That is, masters-level people have the advantage of high-level, specialized training without the detriment of being so narrow so as to restrict the number and types of opportunities available to them. This means that the process of finding a job, as well as of changing jobs, is sometimes easier with a masters degree than with a Ph.D. degree.

The Ph.D. Degree in a Basic Scientific Discipline

If you are looking to direct a research or clinical laboratory, you will need a Ph.D.

However, a Ph.D. is not for everyone, and it will not answer all of your employment problems. A Ph.D.-level person often performs a very specialized function. Of course, some people like to go to school for more education. Certainly, that is a fine reason. However, if you are thinking of the degree as a stepping stone in your career, you may wish to consider what you are planning on doing with it, before embarking on the significant time and energy investment it will take to earn one.

Let's liken a Ph.D.-level worker to a fuel injection system in a car. Is it sophisticated? Yes. Did it take some time to develop? Yes. Does it add value to the car? Yes - usually.

But let's suppose the electric car gains more of a niche in the marketplace. Or let's suppose that public transportation results in less demand for cars in general. Both of these events will drive down demand for fuel injection systems. These situations are obviously not good for manufacturers of fuel injection systems, who are now stuck with an excess of products they can't sell to the car market.

So what can they do now?

•   Nothing. They can eat the losses and move on with the development of other products.

•   Sell the product for some other purpose. Obviously, this alternative depends upon finding another market for the product. It also depends on being able to successfully market the product to the new market.

• Modify the product, and then sell it or its component parts for some other purpose. However, the cost of modification may or may not be worth the potential gains.

Of these choices, the manufacturer will try to pick the most profitable option. Based on this decision, the manufacturer may decide either to discontinue the product entirely or to continue producing it.

When you think about getting your Ph.D. degree, you should probably consider one thing. Higher education means a higher level of specialization. While this can translate into higher-level positions and higher salaries, it can also mean it's harder to find a specialized job. Of course, if you are having trouble finding a job, you can always adapt your skills to a better job market! People with Ph.D. degrees have more options than many people realize. However, for those of you who are just considering entering a Ph.D. program, these are important points to consider.

Other Types of Degrees

Perhaps the type of job you desire requires an additional scientific or medical degree, certification, and/or license. Alternatively, perhaps it involves combining your biology skills with those of a non-biological discipline. There are many different types of degrees which are not strictly for biological research and which can be very marketable in combination with your biology degree and which can provide you with all sorts of interesting job opportunities!

Applied Scientific Degrees

Applied scientific degrees are just that - they are meant to be applied to a particular line of work. For example, a bachelors degree in physical therapy, along with the appropriate licenses and certifications, can allow you to practice physical therapy. Likewise, a masters degrees in genetic counseling, along with the appropriate licenses and certifications, will allow you to become a genetic counselor. There are plenty of applied doctoral-level degrees as well, such as those in human medicine, dentistry, veterinary medicine, and pharmacy, to name a few.

Many applied degrees may be obtained at the bachelors or masters level. For example, you can obtain a bachelors or a masters degree in forensics or medical technology, or you can obtain a masters degree in exercise physiology, and practice in those fields.

If you have already obtained your Ph.D. degree, you can still enter special postdoctoral training programs which can train you for work in an applied field. For example, you can combine a Ph.D. in a scientific discipline with a postdoctoral fellowship which can train you to direct a clinical laboratory. There are also special training programs throughout the federal government for epidemiologists and other applied scientists.

Of course, if you earn an applied degree, you are not always required to work with patients or in the exact field in which you trained. For instance, many pharmacists, medical technologists, medical doctors, exercise physiologists, and so on, may go to work for pharmaceutical or biotechnology industries, other organizations of the private sector, or government organizations.

Degrees in Other Fields

There are also degrees which can allow you to combine your biology skills with those of another profession. For instance, people may combine their biology degrees with degrees in law, computer programming, business, regulatory affairs, education, information and library science, biostatistics, and plenty of others. The combination of knowledge in several different fields can be very valuable in the marketplace.

So, How Much Education Should You Really Get?

When thinking about how much education to pursue, there are many factors to consider. Why do you want to go back to school? Is it for the pleasure of learning? Is it for the prestige? Is it to get a "decent" job?

One thing that you should consider is that education is not necessarily the be-all and end-all of career success. You may have very good reasons for going to school. Nonetheless, it's worth considering whether you need a particular type of degree, *before* embarking on an expensive and extensive degree program.

Financial Implications of Higher Education

One of the first things many of us think of when considering more schooling is the financial implications of it. School often involves a huge expenditure of money, or at the very least, a sacrifice of your earning potential during that time. Of course, money isn't everything. However, since it's an important consideration for many people, it's worth taking a minute to contemplate the financial ramifications of going to school.

We are often told that higher education will result in a more financially rewarding job. However, this is simply NOT always true. Some types of degrees simply pay better than others, and some degree programs take so long that they compromise one's earning potential for a long time.

For example, let's consider three individuals, all of whom obtained their bachelors degrees in biology at the same time in 1988, at 22 years of age. Tom completes his bachelors degree and goes directly into the workforce. He starts in a position earning $22,000 a year.

Dick, on the other hand, decides that he wants to get a Masters of Business Administration (M.B.A.). He pays for 2 years of full-time tuition, at a cost of $15,000 a year. Dick graduates in 2 years and obtains a position earning $47,500 a year (adjusted for 2 years of cost-of-living increases at a rate of 3%. If he had started working with an M.B.A. in 1988, he would've been making $45,000 a year).

Harry decides to obtain his Ph.D. During his schooling, he makes an average of $11,000 a year. He graduates in 6 years. He then completes a three-year postdoctoral fellowship, at an average salary of $23,000 a year. After his postdoc, he finally lands his first position at a salary of $54,800 a year (this figure has been adjusted for 9 years of cost-of-living increase at a rate of 3%. If Harry had had a Ph.D. in 1988 and had started working at the job then, he would have been making $42,000 a year).

Now let's assume that, starting with each person's first position in the "real" world, his salary increases at a rate of 5% a year (cost-of-living increase plus some merit). At age 65, each would have earned in salary the amount of money shown on line 3 of Figure 19.

Figure 19
The Earnings of Three Different Workers

Line	Parameter	Tom (B.S.)	Dick (M.B.A.)	Harry (Ph.D.)
1	Number of Years Working[1]	43	41	34
2	Starting Annual Salary	$22,000	$47,500	$54,800
3	Total Salary Earnings[2]	$3,145,853	$6,072,389	$4,661,669
4	Graduate School Expenses	0	-$42,000	+$135,000
5	Total Earnings Minus Graduate School Expenses	$3,145,853	$6,030,389	$4,796,669
6	Retirement Savings	$435,050	$800,226	$521,564
7	Total Lifetime Earnings	$3,580,903	$6,830,615	$5,318,233

[1]Assuming that everybody retires at age 65.
[2]Assuming that everybody's salary increases at a rate of 5% a year.

Since Dick spent $30,000 in tuition to go to school, we must subtract that amount of money, with interest, from his earnings (Dick probably took out a loan for the money and now must use some of his earnings to pay back his debt). Let's assume that it is an 8% loan, so let's subtract $42,000. Since Harry actually earned money during graduate school, we will add $135,000 to his worth.

Each now has the amount of money shown on Line 5 of Figure 19.

Now let's also consider the fact that most employers provide some sort of retirement benefits to their employees. Let's assume that employers will contribute 6% of an employee's salary to their retirement fund. Let's also assume that these retirement funds increase at a rate of 8% a year. By the time they are 65, Tom, Dick, and Harry have the amounts in their retirement accounts shown on Line 6 of Figure 19.

Adding the earnings and retirement savings together, the net worth of each individual is shown in Line 7 of Figure 19.

Of course, this is a very simplistic model with a lot of assumptions. For instance, it assumes that persons with all degrees obtain promotions to the same extent. In reality, promotions often depend on where you work and how proactive you are in shaping your career. It also assumes that every employer gives the same benefits, and that everyone retires at the same age.

We all know that in real life, there are many more variables than those which have been presented here, which will influence your earnings throughout your career. Certain career paths may afford a person the opportunity to bring in additional income, and certain careers may allow a person to work for a longer time until retirement. The "total earnings" calculation also does not take into consideration the fact that individuals invest differently (some are wiser and luckier than others).

Nevertheless, the model still illustrates one important point. It is not always easy to correlate the level of education with the amount of total earnings over a lifetime of work and pay. The longer you are in school (and hence, not working), the shorter the amount of time that you will have both to earn money and to take advantage of the compounding nature of your investments before you reach retirement.

By the way, these numbers were taken as realistic figures from a few of the author's peers.

Enjoyment of Your Work

Enjoyment of your work is a very important part of many people's decision to go to school. Perhaps the type of work that you would enjoy requires a higher degree. Enjoyment of your work has value, too! You may decide that enjoyment is in fact more important to you than maximizing your financial reward with a particular career path. Thus, you should make it a priority to research the day-to-day activities which will be required of you in particular jobs, and think about how to plan your education so that you can do what you most enjoy.

Promotions

The higher degree is important for promotions in some professions. However, this is not a universal rule - your approach to your career is often at least as important, if not more so, than your education. This concept is illustrated by the fact that presidents and executives of organizations often do not have as much educational training as their underlings. Although these people do not always have a lot of higher education, they have steered their careers in the direction they've wanted them to go. Often, they've also networked and made contacts and have developed impressive resumes which demonstrate their abilities in a particular area.

However, since some career paths do require higher degrees for eventual promotions, you will probably want to research this topic in the career area(s) in which you are interested. Promotions can depend very much on job market, organizational size and philosophy, and many other factors.

Overall Success

What does success mean to you? Is it defined in financial terms, happiness, promotions, or some other factor not mentioned here? What degree do you need to get in order to get the things which are most important to you? These questions are important to consider when deciding whether to pursue more education.

So, How Much Education Should You Get?

Appendix A contains a table listing some examples of professions, and the type and amount of training which is required for them. Please take a minute to locate that table.

While looking at that table, you probably noticed, once again, that there is not always a strict guideline as to how much education you need for a particular career.

The different types of biology degrees, along with the positive and negative aspects of each, have already been mentioned earlier in this chapter. However, for some people considering a combination of a biology degree with that of another discipline, you'll have to decide how much education to get in the other discipline.

 Education Can Be a Gamble!

How Much Education Should You Get in Disciplines Other than Biology?

The factors to consider are pretty much identical to those factors which can influence your decision to go on for further education in biology. There is no right answer.

However, since higher education often doesn't hurt, you may want to go for the highest-level degree program you can get into. For instance, why obtain a second bachelors degree, if you can obtain admission directly into a masters program (which might actually take less time to complete than a second bachelors degree)? Of course, there are some programs (such as computer science) in which you may not be able to be admitted directly into a graduate program in that discipline. You may be lacking too many of the prerequisites. In that case, you may wish to consider getting the bachelors degree in that discipline. Alternatively, you may opt to take only those courses you think would be helpful to you in a job, rather than to enter a formal degree program at all.

A Ph.D. degree may take a little more consideration simply because it often takes so long to complete the program. However, some doctoral degrees don't take as long. For example, the doctor of jurisprudence takes only three years to complete, and a doctorate in pharmacy can take only one to several years beyond a bachelors degree.

In Closing

Education is an investment. Like any other investment, educational investment can be somewhat risky. However, it can also bring you great returns! In any event, your willingness to do your homework *before* making the investment may spare you a lot of expense and may help lead you to a happy and fulfilling career!

Part IV
Stories, Anecdotes, and Personal Examples

Chapter 14
The Journey of a Successful Job Seeker

The following is an anecdote written by a happy worker who wanted to share the secrets of his success:

The lab phone chirped with the double-ring characteristic of an off-campus call. "Hello", I answered. "Dr. Mad Scientist's lab. This is Pete the Postdoc."

The caller identified himself as the Human Resources representative for a company at which I had just interviewed. "I have some good news for you", he said. "We would like to offer you a job as a medical writer in our company." The conversation then turned toward the topics of salary and start dates. The call ended with my agreeing to contact him within the next five days.

I hung up the phone and screamed with joy. I had finally gotten a job offer that I wanted. After several months of no offers, I was beginning to doubt whether I'd ever make the transition. But here I was now, looking forward to financial solvency, better working conditions, *and* the opportunity to get some good work experience!

I thought about how far I'd come since I first began to really think about getting a job. This had happened late in my undergraduate training, when I was trying to decide about whether or not to go to graduate school. I had drawn up a flowchart of what I perceived to be my different career options, considering what might happen if I ventured off to graduate school versus the "real" world of work. It is reproduced in Figure 20 (next page).

Having decided that graduate school was in fact my best (or at least the most readily available) option, I found myself five years and thousands of cell cultures later, still not knowing what I wanted to do. I had been working at a fanatical pace, desperately trying to generate enough data for a thesis and some manuscripts. I didn't want to think about the fact that these materials would probably be read by an audience of three, even including the five individuals who would serve on my Ph.D. thesis committee.

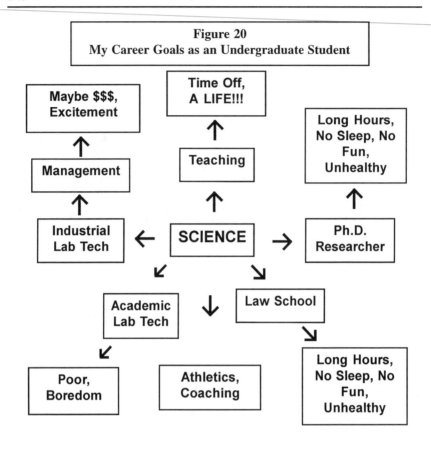

Figure 20
My Career Goals as an Undergraduate Student

My fatigue and hunger of one particular evening during graduate school had particularly intensified my growing apprehension about my job search. I was nearing the end of my program, and I needed to find a career. There I was at eight o'clock in the evening, yet again a prisoner to my experiment, waiting for the scintillation counter to produce the results for which I had devoted my entire week. Somehow the long hours in the lab didn't seem to be very effective in helping me to find a job.

Right then, I knew that it was time to seriously explore my career options. Not only did I need to find a way to increase the measly figure reported on line 1 of my 1040EZ, which was seriously deficient in relation to my expenses, but I realized how badly I need a perspective when I thought about claiming *Escherichia coli* as a dependent on my tax return. True, it had been the only organism I had been cultivating for the last five years. Still, though, I didn't think this would fly with the IRS.

I began to think about what I wanted out of my life. I realized that my primary goal, corny and nebulous though it was, was to find happiness. Career satisfaction was one component of that formula. It consisted of the need to balance the sometimes conflicting goals of earning money, enjoying my work, and keeping decent working hours. Although I did not know how to achieve my long-term goals at this particular moment, I believed that somehow, some way, it could be done.

Not knowing what else to do, I updated my flowchart with what I now perceived my career options to be after graduate school. It is reproduced here in Figure 21.

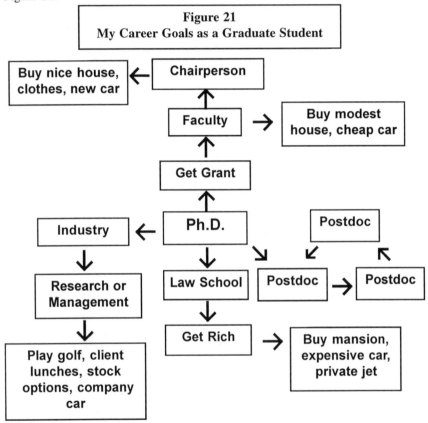

Figure 21
My Career Goals as a Graduate Student

Did you notice that swirling, vicious circle-type figure on the right side of Figure 21? This was the path that most students took after graduating with their Ph.D. degrees. It looked to me sort of like a whirlpool. Having done some kayaking in my undergraduate years, I knew how difficult it could be to get out of one of those things. By analogy, I began to call this particular career path the "postdoc whirlpool".

I knew I didn't want to fall into this trap. The pay was poor and the hours were long. Somehow it didn't seem consistent with my newly-stated goals in life. Besides, even though I had learned to tolerate, and at times even like, bench-level research, I was not convinced that I wanted to keep doing it for the rest of my life.

One thing I couldn't help but notice, in comparing the two figures, was how the endpoints of these charts had shifted from idealism to finances. Is that what years of graduate student poverty had done to me?

Anyway, after studying these charts for a while, I realized that I needed a much better picture of what was really out there. These charts had not been very well-researched, and here I was trying to make a career decision based on them! Somehow this did not seem like the most effective approach for getting a job that I wanted. I decided that a change in tactics might be in order.

I then had an idea. I would approach my job search much like I approached a scientific problem. I would come up with a hypothesis, do the necessary background reading, gather data, and make conclusions. I would also make note of future experiments which should be done, and areas in which further information was needed.

The hypothesis was easy: Somewhere out there, there was a satisfactory job for me. Once I had the hypothesis, it was the research that turned out to be the challenge.

My research went along like this:

I first approached some of the people I knew best - the faculty at my university. I asked them about the ways of the world and the types of job opportunities in it. It did not take me long to figure out that asking university faculty about careers was like calling a plumber to fix an electrical problem in my house. The conversation usually went along something like this:

<u>Pete:</u>   "I am interested in exploring career opportunities both at and away from the lab bench. How much do you know about the different types of career opportunities that exist for scientists?"

<u>Prof:</u>   "Well, basically, you can do bench-level research at pharmaceutical and biotechnology companies, as well as in academia. The government offers opportunities as well. It depends on what you want - in a company, you may not be able to work on the projects you want to, but on the flip side, you don't have to apply for grant money. In the government, you may have responsibilities outside of research, which may interfere with your ability to do research."

<u>Me:</u>   "What types of responsibilities? As I said, I am interested in the different types of work there are to do, not just bench research. Although I like bench research, sometimes I find it rather confining and narrow. I'd like to expand my horizons, and learning about these opportunities may help me to do that."

<u>Prof:</u>   "Well, for instance, at the FDA, researchers often have to do some reading of New Drug Applications. It can really take a lot of time. They also have to spend time in meetings reviewing these applications. It all takes away from their research time."

<u>Me:</u>   "It sounds like interesting work, though. It is a way to see what is

happening with the world, and where all this research eventually leads."

Prof: "Yes, that's true, but the primary function of a Ph.D. is to perform research."

Me: "Aren't there other types of opportunities for a Ph.D.-level scientist? I mean, research is interesting, but I would think that there are other things that a Ph.D.-level scientist is qualified to do."

Prof: "Why did you get your Ph.D. if you don't want to perform bench-level research? That is what Ph.D.s are trained to do. If you didn't want to do it, you should've gotten some other type of degree."

Me: "Thanks very much for your time."

It was time to nip that conversation in the bud. Sure, these conversations might help me to learn a little about other opportunities, such as doing review of applications at the FDA. Nevertheless, I'd be retired before I'd learn about the different opportunities at that rate.

I did, however, learn about a few publications that might be interesting to me.

Next I went to my university's career center. I asked the receptionist whether they had any career books written specifically for biologists. She pointed me to a shelf. I thumbed through the books, but I couldn't find any which contained the type of information I was looking for. I scheduled an appointment with a career counselor.

Me: "I'm here to talk about the career opportunities available to me in the working world."

Career Counselor: "Do you know what type of work you would like?"

Me: "Not really. That's partly why I'm here. How do I find out about different opportunities? I feel like I don't even know where to start."

Career Counselor: "Just start calling people. Get names, if possible. Then get on the phone and ask them questions about their work. You'd be surprised at how people will talk to you. If you don't know a name, concoct a story so that you can get one."

Me: "Like what?" It wasn't my most creative moment.

Career Counselor: "Well, I know people who've said that they're doing a story on such and such, and that they would really like to talk to someone in a certain position. They then ask who might be willing to talk to them. You can think of all sorts of devious schemes. It can be fun, and it works."

Hmm, I thought. This wasn't my usual personality. I didn't normally have that kind of courage. But what the heck, I needed a job, and if I really got desperate, I might consider trying it! I would use this approach as a last resort, kind of like I would parachute-jump out of a plane if I knew it was going to crash.

In the meantime, the parallels between this career search and my bench research were becoming increasingly apparent to me. The progress was slow, and ninety percent of the time I felt like I was getting nowhere. "Research" was increasingly becoming an eight-letter word to me - it was twice as bad as any four-letter word!

Soon after that conversation with the career counselor, I spotted a sign at my university which announced a "career counseling" session. This session, which was intended for M.S. and Ph.D. graduate students, featured alumni of the university who were now working elsewhere. Each alumnus was to talk a little about his or her career. Each alumnus was also available to answer questions and to talk to interested students one-on-one. I was really psyched - finally, here might be my chance to learn about career options outside of the university!

Disappointingly, the session started slow.

Alumnus: "Hi, I'm Dr. Mark Mywords from Frenzied State University. My lab is involved in researching the genomic organization of *Bacillus subtilis*. In some isolates of this organism, the grw gene is disrupted by a bacteriophage. A functioning grw gene controls the growth rate of the organism; therefore we'd expect that organisms which are entirely grw- would probably grow more slowly than those which are grw+. In fact, we find that both genotypes have the same growth rate. My lab is trying to understand how the organism adapts to the lack of a grw gene. We are currently the leading researchers in this field worldwide."

Me: "Zzzzz."

Big deal. I wondered why I'd come. I wanted to hear about careers, not a research seminar.

As desperation gave way to boredom, I seriously considered leaving the session. Then I reconsidered. Knowing my luck, someone interesting would speak as soon as I left. I decided to hang in there a little longer.

I was really glad I stayed. It turns out that there was one person whose career particularly interested me. This person had abandoned basic research and a university professorship in favor of an "alternative career" doing clinical research in a pharmaceutical company. He talked about his position as an associate director of a therapeutic unit. He also mentioned other positions such as medical writing, drug safety surveillance, and clinical monitoring.

I approached this alumnus after the session.

Me: "Thank you for speaking today. I enjoyed your talk. I'm interested in learning more about careers in the pharmaceutical industry. For instance, I'd like to know more about medical writing. What is it, and what experience would I need to have in order to do it?"

Alumnus: "Medical writing basically involves writing about clinical trials of pharmaceutical agents. You would work in conjunction with biostatisticians to analyze and make sense of clinical data. The pharmaceutical industry is always looking for medical writers."

Me: "Might I be qualified for such a position? I will be receiving my Ph.D. in microbiology soon."

Alumnus: "Yes."

Me: "How can I learn more, and how can I break into the field?"

Alumnus: "Here's my business card; please give me a call and we can talk some more."

Wow. Now I was both intrigued and excited. But I was still perplexed over one thing: If these jobs were prevalent, why had I never heard of them before? My answer to myself was simple. I had never been out in the world, so how would I know?

I now saw that the university's career counselor had had a good point. I needed to start talking to people. Granted, her approach of calling people would require somewhat of a personality transformation for me, requiring me to change from that of a hermit-like researcher to an outgoing job seeker and back again. Nonetheless, I decided that this feeling was probably better than never getting a decent job.

In the meantime, the alumnus I had just met had agreed to host me as part of an "alumni shadowing" program run by my university. This program was designed so that active students could tag along with alumni at their workplace, to see the types of work that alumni did.

When I arrived at my newly-found ally's house the night before I was scheduled to accompany him to work, I could tell that it was going to be a very productive visit.

Alumnus: "The pharmaceutical development process occurs in stages. Potential new agents are first discovered and tested in the lab, then in animals, and finally, in humans. Each one of these stages has its own tasks that must be performed. For this reason, in order to understand the career opportunities in the pharmaceutical industry, it is very helpful to understand the pharmaceutical development process in general. Let's start by going over the basics."

As the alumnus described the process to me, I tried to think about how the different aspects of the development process could translate into job opportunities. I could see that there were a lot of different tasks to be done, each of which would require a specific type of expertise.

The next day, my new ally had set up appointments for me to talk with someone in regulatory affairs and someone in project management, in addition to himself. He wanted to give me an idea of the diversity of tasks that existed in the pharmaceutical industry. The pharmaceutical industry in fact seemed to be a goldmine of career opportunities!

When I returned to the alumnus' office after these appointments, he handed me a piece of paper with a long list of names.

Alumnus: "I've been writing down some additional information that might be useful for you. As I mentioned before, the pharmaceutical industry sometimes outsources some of its work to contract research organizations, also known as CROs. Since it is difficult for someone who's not in the industry to know how to identify good CROs, I've listed some of the more reputable ones here. You may wish to look for a job at one of these, since sometimes it is easier to land your first job in a CRO than in a pharmaceutical company. The pay is a little lower and they'll work you harder, but it will also be great experience."

I left with a feeling that I was very familiar with, especially during my early years of graduate school. I felt overwhelmed. There was so much I didn't know, and so much information to absorb and integrate.

Nevertheless, this wonderful alumnus had started me down the road. He had been so very helpful to me, I couldn't thank him enough. Before I left, he also had written down the names of some relevant professional societies and had also given me the names of some people to call.

Now armed with this information and the support and encouragement of my friends and the career counselor at my university, I got on the phone and started calling people. When I depleted the contacts that my pharmaceutical-worker ally had given me, I began to search for more names from the directories of relevant professional societies. I also tried to do that "networking" thing that everyone talked about.

Me: "Hi, my name is Pete the Postdoc. I am interested in positions in medical writing, and I noticed that you are involved in it. I am wondering whether you would be willing to talk with me about your experiences in the field."

Most people were surprisingly accommodating.

Worker: "Sure, what would you like to know?"

Me: "I am wondering how you got involved with it, how you like it, and what it entails on a day-to-day basis."

As I might have expected, the answers varied. Many people had either started in the laboratory or in a clinical profession. Of the ones who had come from the laboratory, some of them missed it and some did not. But no one disputed the fact that the pay and the job market were better in medical writing than in the laboratory.

After talking to a number of different people, none of whom expressed an unconditional regret about leaving bench research to go into medical writing, I decided I wanted to give it a try.

<u>Me:</u> "Do you know of any positions which are available, or do you have any advice for a person trying to break into the field?"

<u>Worker:</u> "I am not sure whether our company is planning on hiring at the moment, but why don't you submit your resume to so-and-so, or give so-and-so a call? You can tell them that I referred you to them. Also, sometimes I get calls from recruiters. Would you like me to pass your name on to them?"

<u>Me:</u> "Why, sure, that'd be great. Thanks so much for your time and help."

I was so amazed. These people were so helpful. Before long, I was starting to receive calls from recruiters who had gotten my name from my new contacts. Unfortunately, I found out that many recruiters don't deal with "entry-level" people.

Eventually, though, the calls did pay off. I had managed to directly identify quite a few open positions. Although most companies were looking for people with prior medical writing experience, I was able to sell a few companies on the fact that I had in fact done a lot of writing throughout my doctoral work and that I would be a motivated employee.

And so I received my first job offer.

Sadly yet elatedly, I packed up to go. I was very excited to begin my new life. Yet at the same time, the moment was sad simply because I was now leaving the lab where I had worked for more years than I wished to count!

As hard as it was to say good-bye, I love my new life. I'm learning a lot in my new job. I'm also making more money than I would've ever envisioned several years ago.

Throughout my job search, though, I kept wondering why there were no books to provide me with a good overview of the working world and the jobs within it. It sure would've made my job search a lot easier. It all worked out in the end, but it sure was a struggle!

I know that this book would've been helpful to me when I was looking for my first "real" job. I hope it's helpful to you, too.

Chapter 15
A Day in the Life of Seven Young Workers

Contents
A Day in the Life of:

*Please note: These summaries have been obtained from people who work in these organizations; however, names and some details may have been changed.

A Day in the Life of:

Laura Lab, Ph.D.
Research Scientist
A Pharmaceutical Company

Description of employer and position: I work in a pharmaceutical company that develops drugs to treat disorders of the central nervous system in people. I work in preclinical research. My responsibilities are to work on project teams designed to identify compounds of interest, to test new compounds developed by chemists, and to contribute new ideas for therapies in the central nervous system.

Daily activities: No one day of activities is really typical for me. This is because the experiments which I perform are often long and time consuming. Therefore, I must either be doing work in the lab all day or not. Every week I spend about 3 full days doing experiments. The other 2 days are spent reading, analyzing results, writing, and doing administrative chores. The project teams I work on meet monthly and I attend those meetings and present my progress. Typically I would have another person working with me, helping me to prepare my experiments and run the lab.

Other tasks: We write exploratory research proposals which are reviewed twice a year. If a potential project is deemed attractive, resources are allocated to do that proposed work. We also present progress to management committees which evaluate the progress of each ongoing project and determine when to move something on to the next step or when to discontinue it. I travel to 2-3 meetings per year and establish contacts with outside scientists. I can present non-proprietary work at meetings and I can publish this work in peer-reviewed journals.

Aspects of my job that I like: I enjoy the pace of working in industry. In my experience the research goes a bit faster than in academia due to issues around patent life and such. Also, I find that industry has a much more team-oriented environment. I like working together with others who have areas of expertise very different from my own, like chemists and toxicologists. This gives me more of the "big picture" and keeps the work interesting. Finally, grant writing is a necessary task in academia. In industry, we write proposals and make presentations, but on average I think we spend less time pleading for resources. Money flows more freely. Time is the more valuable commodity.

Aspects of my job that I do not like: Because this is a business, and because there are several levels of management who are somewhat removed from the actual work being done, it is necessary to give them frequent reports of what we are doing and how we spend our time. Often this takes a bit of time in itself and takes time away from getting experiments done in the lab.

Other comments: Generally, once something is classified a company priority, things move quickly and money flows freely. Sometimes one must pass over interesting details of an experiment in order to stay focused on the goal.

People skills are a plus in industry, probably more so than they are in academia. There are more opportunities to interact with other scientists and to market oneself. In the academic environment, with enough hard work, creativity, and genius, one can become successful --- regardless of one's interpersonal skills. My impression is that, on average, industry scientists are more well-rounded as people and have more of a life outside their work.

How to get a job like mine: A surprisingly large percentage of people get jobs like mine through a personal contact. Industry is no different. Competition for jobs is pretty heavy. Some positions are not advertised. Some are filled using private agencies. Often positions are created to accommodate someone with exceptional skills, who has a contact.

One way to get one's foot in the door of industry is to try a posdoc, co-op, or temporary job in industry. Many times the individuals in these situations demonstrate their skills and the significance of their work to management and management does not want to lose them. They may get hired-on into "permanent" positions. Of course this is not always the case. For this reason it is critical, especially for posdoc fellows, to keep an eye on the job market and on their marketability (bolster that curriculum vitae!). Do not fall into the trap of working on proprietary projects without the possibility of publishing, unless you have a permanent position! Some companies have employment policies against hiring their posdocs into regular jobs. This can be a good situation, because it ensures that you will be leaving and your goal is clear to everyone---strengthening your CV. In addition, the industry posdoc should help to make available more options for you. You can now choose what you prefer: the academic environment or the industrial one.

A Day in the Life of:

Vince the Validator, M.S.
Systems Coordinator
A Biotechnology Company

Description of employer and position: My company manufactures a contained polymerase chain reaction pouch which tests for the presence of various infectious diseases. The DNA of an infecting virus can be amplified and the presence or absence of an infection can be known in 2-3 hours time from sample preparation.

As a systems coordinator, I oversee the validation activity that is going on in each unit of the factory. Such validations can be initiated internally or can be set up with the help of an outside validation firm. If such a firm is involved, I serve as the focal point between the validation firm and the employees at the manufacturing company. Essentially, I communicate technical information between groups to ensure that the validations are being completed in a timely fashion.

Daily activities:

Early Morning: Answer voicemails and e-mails.
Generally, the questions are related to the various validation projects that I am involved in. I often make arrangements for outside validation firms to come in and work at our company. This involves checking schedules.

Mid-Morning: Plan Projects and Work on Miscellaneous Tasks.
I may write Standard Operating Procedures related to validation projects, major utilities, or common pieces of laboratory equipment. I am planning the scope of future validation projects.

Early to Mid Afternoon: Collect any validation documents and forward them on to the Director of Compliance. I may also spend time updating each group's validation progress. I often write up meeting notes for general distribution.

Other Tasks: I often arrange meetings to discuss validation issues. I need to know which technical experts are needed for each project and contact them. I also present the status of the various projects to the validation team and to management. I have attended conferences related to computer systems and methods validation. I often serve as a consultant for other groups such as the Information Management group, to determine if their computer system validation efforts are comprehensive enough to satisfy Food and Drug Administration (FDA) Guidelines.

I have also provided training for groups which consists of presenting prepared training modules on overheads.

Aspects of my job that I like: I like interacting with people from different backgrounds. I am used to dealing with biological scientists but in this position, I deal with engineers and other technologists.

I like being involved in several projects at once. I enjoy the people I interact with.

As a communicator of technical information, I enjoy communicating with people in writing, as presentations to groups, and/or on an individual to individual basis.

Aspects of my job that I do not like: I wish I had more time to become more deeply involved in projects. I miss the analytical problem solving skills that I am accustomed to using as a biological scientist.

Oftentimes, I feel spread too thin because I am involved in many projects in a fast paced environment. Sometimes, it is not obvious which projects take priority. This is a sign of our organization more than my job.

Other comments: The validation job market is good but is very competitive for small, outside validation firms. The pay is rather good for this position but in return, longer than the standard 40-hour workweek is required at times.

How to get this type of job: Generally, people in this type of position have an advanced degree in the sciences or in engineering. They often have a wide range of experiences within their fields and are proven problem solvers. One should have above average to excellent writing skills. I got into this job from a method development position in which I transferred a protein identification method into a manufacturing laboratory. I learned about a whole new field - HPLC (size exclusion chromatography and ion-exchange chromatography). I had to qualify the HPLC system and run studies to transfer and validate the method. I showed a willingness to tackle projects that were unfamiliar to me and to become an expert in the area. I also trained several operators on how to performance check the equipment and run the method. I wrote comprehensive documents that covered standard operating procedures (SOPs) for performance checking and maintaining the equipment. I also wrote SOPs for running the methods, preparing samples and interpreting data. I wrote the complete equipment qualification package as well.

I am glad that I have been able to combine my scientific training, writing skills, and communication skills to complete validation and development projects in a timely and competent manner.

A Day in the Life of:

Rick the Reviewer, Ph.D.
Microbiologist
U.S. Federal Government

Description of Employer and Position: I work for a government agency which promotes the development of new and useful biological products. My agency monitors the purity, potency, safety, effectiveness, and consistency of these products. To this end, we propose and enforce regulations designed to promote the safety of potential consumers of these products. We review applications for licenses for biological products from private industry and inspect their manufacturing facilities.

My particular position involves the review of applications and the inspection of manufacturing facilities for biological products. I typically work with 15-20 different organizations at any one time.

Daily Activities:
8:00 - 8:30 Read incoming e-mail messages, listen to voice mail messages
Messages can be from other workers within my agency or from outside organizations.

8:30-12:00 Read and review applications
If possible, I use this time to read applications from manufacturers of biological products. Otherwise, I use this time to make phone calls and respond to e-mail messages, or to attend meetings with coworkers or with outside organizations.

12:00-12:30 Lunch

12:30-2:30 Meetings
Meetings can either be internal (for instance, to discuss an application or an organization), or external (for example, with an applicant). They may be conducted either face-to-face or as a teleconference.

2:30-4:30 Respond to messages, perform administrative tasks
I use this time to answer e-mails, return phone calls, and to arrange other necessary meetings and phone calls. I also spend time writing meeting minutes and documenting phone conversations. I make travel arrangements for upcoming inspections.

3:30-6:30 More review of applications
Again, I try to find some time to read and review applications. Since this time is often interrupted by phone calls, meetings, administrative tasks, and other urgent issues, I stay as late as is necessary to finish the work according to required timelines.

Other tasks: My position involves the inspection of manufacturing facilities, as well as the review of manufacturing procedures. I therefore travel approximately 50% of the time. Inspections are very important components of the overall review process, since approval to manufacture is only granted for very specific physical locations (i.e., specific rooms and buildings), using very specific manufacturing procedures. Any changes to the licensed physical location or to the licensed process must be approved, generating another round of reviews and inspections.

Currently I do not have my own research projects - I do not have time! However, time permitting, I could become involved in a research project.

Aspects of my job that I like: I enjoy reading about the different products, doing reviews and inspections, and helping organizations to solve problems. I use my scientific training daily. I also enjoy the travel, through which I've had the opportunity to meet many interesting people. I travel to many interesting places - for instance, I've had 3 foreign trips this year and already have 4 planned for next year, to inspect foreign facilities for approval of their products within the U.S. I very much enjoy the collaborative nature of my job, which involves both industry and other people within my agency.

Aspects of my job that I do not like: There is a lot of pressure within my organization to stay on timelines and to act quickly. I also realize that I must be very careful about everything I say and recommend to a company, because it must reflect all current regulations and agency policies. We cannot talk to the press, but companies can. Sometimes this is frustrating, because companies are free to provide their own perspective on an issue, and we cannot respond. This can make us look bad in the public eye and create additional pressure within the organization.

Other comments: I enjoy my job and feel that my scientific training has prepared me well for the continuous challenges of my work. The job market for my type of position is excellent once a person has a few years of experience. Many vacancies are created because people often move on to industry after a few years at the agency. It is also easy to move to different positions within the agency. It is possible to do short (1- to 3-month) details at different jobs to see what other positions are like.

The salary is comfortable - starting salaries are probably around the mid-to high-$30s for someone with a bachelors degree, and a little more for someone with a Ph.D. This job offers a lot of stability and security. It also provides a lot of diversity - for instance, some of our employees teach college at night, and others do research. We often receive requests to give talks at government, industry, or other meetings. I give several lectures each year in graduate classes at area universities.

How to get a job like mine: I learned about this job from a personal contact within my agency. This person faxed me the job description for my current position and encouraged me to apply. Although all jobs are posted publicly, personal contacts can help a person to identify and find positions since it helps to know where to look for the jobs.

A Ph.D. is not required for my position, although I do find that my training helps me to evaluate applications and problems. I also think that my advanced degree gives me credibility in others' eyes when I make recommendations.

Just about everyone in my type of position has at least several years of experience in a research laboratory. Other skills which are very important to this job include the ability to work with others and to communicate well. For instance, it is important to be able to think before speaking and to communicate clearly, because industry often makes long-term, expensive decisions based on our recommendations. Recommendations must be solid and well-supported. It is essential to be able to defend one's own opinion about a particular decision, or how a situation should be handled. Consensus building is critical. Good writing ability is essential for the generation of reports and meeting minutes, and for documenting important conversations with others.

A Day in the Life of:

Scientific Susie, Ph.D.
Research Associate
A State Legislative Commission

Description of Employer and Position: I work for an in-house state legislative commission on science and technology whose purpose is to guide the legislature through issues involving technical and scientific legislation. In my position, I am responsible for identifying important scientific and technical issues, preparing reports about them, and drafting legislation based on the reports. These reports are intended mainly for legislators, although they are also available to the general public. Their purpose is to provide legislators with technical and scientific background behind certain legislative bills.

Because these reports are large, comprehensive, and require a lot of work, an idea is presented to the speaker's office to assess whether there is a general interest in the topic prior to beginning work on the report.

In addition, from time to time, I am asked by a committee of the legislature to review bills introduced in the legislature or certain agency programs and provide an evaluation.

Daily Activities:

8:30-8:45 Answer voicemails and e-mails.
Calls may be internal and administrative in nature, or they may be from legislators or political staff who have questions about a report or a scientific bill. Occasionally, there may be a request for me to investigate a government agency's methods for implementing a particular policy or performing a particular scientific activity.

8:45-12:00 Work on reports and/or legislative bills.
Most of my time is spent researching, reading, thinking, sending for articles, and writing comprehensive reports on scientific policy and legislative issues. Depending on the stage a particular report is in, the activities may be somewhat different. In the beginning, I spend time reading and doing research to identify important scientific policy issues. Once I identify a topic, the bulk of my time is spent researching the topic and writing a comprehensive report about it. These reports can take many months to complete.

In order to prepare my reports, I read scientific and other articles on the topic (for example, ethical or philosophical articles and legal commentaries). I look for model legislation in other states. I then make recommendations about a model bill, or sometimes I draft the bill myself.

Every so often, I am asked to do more short-term work such as review a legislative bill and provide comments on it. In other cases, I may be asked to investigate and oversee a state agency's policies and methods of operation. In those cases, I may visit the agency, research what they are doing, and talk with them on the phone. I question them about the logic and reasoning behind their operations and analyze the appropriateness of their methods for addressing certain policy issues. I then prepare a memo or report about the agency and my findings and opinions regarding it.

These interim assignments, memos and reports usually are much shorter than my usual reports and take 1-3 weeks to complete. My other reports are always ongoing and consume the majority of my time.

12:00-12:30 Lunch.

12:30-5:30 Continue working on my reports.
I return to working on my reports. In most cases, this involves more reading and writing, although sometimes it involves following up with others. For example, I may need to answer questions from legislative staff or meet with them to discuss a bill or report I am working on. I also keep current with events in the legislature.

Other Tasks: Because my reports go through stages from the initial idea to the preparation of a report and sometimes to the drafting of a bill, my day-to-day activities can be quite different from that which is outlined above. For example, once a legislative bill is drafted, I spend considerable time talking with relevant legislative staff to answer their questions and concerns, and sometimes to make them aware of issues and to try to convince them of the importance of the bill. Sometimes bills are passed in one session, and sometimes they can be ongoing and unresolved for a number of years. The extent of my involvement at any one time can vary considerably, depending on what is happening in the legislature at any one time, and what is coming up in the legislature.

From time to time, our group may also organize a conference on a particular topic. We organize the conference and get speakers for it. We do this in order to promote awareness of particular policy and legislative issues. We also hold public hearings on certain policy issues or on a particular bill before the legislature.

Aspects of my job that I like: I enjoy reading and preparing reports about scientific issues. I also like the fact that my work can make a difference - when a bill is passed, it affects the entire state! There is a tremendous sense of personal pride and accomplishment in identifying an issue and being able to effect a legislative change regarding a particular scientific issue.

Aspects of my job that I do not like: Sometimes I am frustrated by the political nature of the job. Politicians are not always in search of truth; instead, they are more interested in getting re-elected. Thus, sometimes it is impossible to effect a change simply because an opposing group has a very strong lobby.

I also find that without political connections, it is hard to advance in my job, despite the fact that I have more scientific knowledge and education than my superiors. This can be a source of frustration to me.

Other comments: Unfortunately for now, the job market for my type of position is not very good. The reason for this is because there are not many legislatures that have a scientific and technical commission. Often legislatures simply contract out this type of work on an as-needed basis, rather than hire full-time staff. Some individuals are recognizing the increasing need for these types of commissions; however, there may be strong political pressure not to have them. For instance, some groups with a very strong lobby may not like the findings and reports of a scientific and technical commission.

All of this may change, however, especially as science and technology continue to advance.

The salary for this type of job is not very high. Starting salary for someone with a masters degree in my type of job is in the low- to mid- $20s.

How to get a job like mine: I found out about my job through a personal connection. For those who do not have connection, however, every commission like mine has internship opportunities for undergraduates. Graduate students and postdoctoral fellows may also be able to do internships. The legislatures will have internship offices. You can find the phone number for the legislature of any state capital, call the public information number, and ask for their internship office and/or opportunities.

If you call the legislature of a state, you can also ask them for their web page address, which will have information about the happenings of the legislature. You can learn a lot about the kinds of things a particular legislature is doing by looking on the Internet. You can find out what's being debated and look through minutes of past meetings. All of this information is public knowledge.

Legislatures also produce annual books of their organization and activities. If a particular legislature has scientific and technology commissions which are established, they will be in legislature's annual book. Additionally, legislatures print daily calendars of their activities.

A Day in the Life of:

Pipetting Pete, Ph.D.
Research Assistant Professor
An Academic Institution

Description of Employer and Position: I work in a university with emphasis on biomedical research and medical education. As a Research Assistant Professor, my primary role is to run a funded, productive research lab, which includes grant and manuscript preparation, supervision of lab employees and students, conducting laboratory experiments, and coordination with collaborators at other institutions in both the U.S. and Europe. Although my official employer is an academic institution, the funding for my salary and the salaries of my employees (as well as laboratory supplies) comes from the National Institutes of Health and other granting agencies. Although it is not part of my job description, I volunteer to give occasional lectures to medical and graduate students, and participate in interviewing applicants to the university.

Daily Activities:

8:45-9:00 Read e-mail messages, listen to voice mail messages.
Messages are usually from collaborating scientists with technical questions and updates on projects. I also subscribe to an e-mail service which informs me of funding opportunities in my research area. Some e-mail messages are internal reminders about faculty meetings, upcoming seminars. I try to answer the e-mail messages at this point, as well.

9:00-10:00 Attend to cell cultures.
At least 3 days a week, I maintain cell cultures in our satellite tissue culture laboratory. I set up experiments, assist my students with their cell cultures, and maintain the equipment (incubators, laminar flow hoods, liquid nitrogen tanks) necessary to carry out experiments.

10:00- 11:30 Work on manuscript(s) or grant(s) in progress.
On any particular day, I am usually working on at least one grant and/or manuscript, sometimes more. This involves researching library references, composing text at the computer, cutting and pasting photographs, and incorporating comments from co-authors.

11:30-12:00 Lunch
I often "surf the net" during lunch, and check/answer more e-mail messages.

12:00-1:00 Meet with lab employees and students.
This can occur informally and sporadically during the day, as my employees and students are all part-time. I usually try to set aside a significant amount of time

for each person for data analysis, planning for future experiments, and administrative duties (performance evaluations, scheduling, etc.). This is an extremely important part of the job, as it keeps focus, continuity, and direction to the lab.

1:00-3:00 Laboratory experiments.
This can involve microscopic observations/photography, cell counting, running gels, etc. This is categorized as "benchtop", hands-on work. The amount of time devoted to experiments varies from day to day, and can be pre-empted if an important grant deadline approaches.

3:00-4:15 Work on manuscripts/grants in progress.
I continue with the work described in the 10:00 timeslot. By this time, my part-time staff has left for the day, and the lab is quiet for this kind of work. I am also ready to return to this rather intense work, after breaking up the day with the activities described between 10:00 and 3:00.

4:15-4:45 Order laboratory supplies, perform grant accounting.
I prepare and phone-in orders for laboratory supplies and equipment. I maintain a Quicken file in my computer which keeps track of the 3 current grant accounts. Each grant account deals with a specific project, so it is important that I organize the items to be ordered accordingly.

Other Tasks: I attend monthly faculty meetings to keep up with current policies and departmental business. I also attend and participate in a monthly journal club in the Ophthalmology Department, in which a recent scientific paper is presented. I make regular visits (at least monthly) to the medical library to keep up with the scientific literature. Once per year, I attend and present a poster or slide presentation at an international scientific conference. I have to prepare annual reports for my chairman, as well as yearly progress reports for funded grants. I write letters of recommendation for current and former students, and review manuscripts for peer-reviewed journals.

Aspects of my job that I like: I love the freedom and flexibility of my schedule, which is almost entirely at my own discretion. I feel my lab is truly my own. I enjoy participating in the progress of diverse research projects, and the opportunity to handle both bench work, as well as administrative duties. I like the stimulating interactions with my lab employees, as well as with interesting colleagues all over the world.

Aspects of my job that I do not like: The lack of job security can be maddening. My position is entirely dependent on grant funding— no grant, no job. I do not like the fact that grant funding is so competitive and time-consuming to obtain. I do not like the university's attitude that internal candidates should not be promoted. My salary leaves a lot to be desired, at present.

Other Comments: I enjoy my job immensely. My lab has been incredibly productive, and morale is good. Although the salary and job security are poor, there is great opportunity for creativity, independence, and collaborations. The job market for the next promotion (Assistant Professor) is very tight. I keep a watchful eye in the job positions posted in *Science*, but most are incredibly specific, or would require a significant geographical relocation, which I am not willing to do, at present.

How to get a job like mine: A "Research Assistant Professor" requires a Ph.D in a scientific discipline, and normally at least 3 years as a postdoctoral fellow in a faculty member's laboratory. It is unusual for a "Research Assistant Professor" to have an entirely independent lab, as I do. Generally, at this stage, one is still associated with a more senior faculty member. I ended up in this independent position after the tragic death of my postdoctoral advisor. However, most Research Assistant Professors do have a great degree of independence. Non-tenure track faculty positions, such as mine, are posted in scientific journals. Another way to find this type of position is by networking at scientific conferences, word-of-mouth from respected faculty members, and contacting authors of interesting scientific papers. I was referred to my initial postdoctoral position by consulting with a faculty member for whom I had volunteered as a laboratory assistant during one summer of my undergraduate life.

A Day in the Life of:

Careful Carol, Ph.D.
Associate Clinical Research Scientist
A Contract Company

Description of Employer and Position: I work for a full-service contract research organization which provides support to the pharmaceutical and biotechnology industries. As an Associate Clinical Research Scientist, I have responsibilities both as a medical writer and a project manager. As a medical writer, I am responsible for writing final study reports, manuscripts, and parts of New Drug Applications for submission to the Food and Drug Administration. As a project manager, I am responsible for managing projects with regard to communication with clients and coordinating the in-house work to ensure that it is performed on time and on budget. I often work on several contracts at any one time, although the number varies depending on the workload of each particular contract. Each individual contract between my company and its clients is unique with regard to the scope of services provided and the amount of work that is required.

Daily Activities:

8:30-8:45 Read incoming e-mail messages, listen to voice mail messages.
For my medical writing responsibilities, messages are usually questions or answers from biostatisticians who are performing the data manipulation to produce tables and figures for reports. For my project management responsibilities, messages may include questions or answers from the client or from team members within my company.

8:45-12:30 Organize and work on my various projects.
The project(s) that I choose to work on during this time are the ones which need the most attention. Sometimes it is hard to decide which projects to work on. I may try to do some medical writing at this time, particularly if a report is due sometime in the near future. However, if my project management responsibilities are too urgent to wait, I will attend to those instead (making calls, sending faxes, writing letters, project planning, etc.).

12:30-1:30 Lunch

1:30-2:00 Answer correspondence (e-mails and voice mails), prepare for team meeting.
I look through and file hardcopy and e-mail documentation of correspondence between my company and its clients, as well as internal correspondence and action within my company. As a project manager, I pay particular attention to client requests and things that will affect the timelines and/or budgets of the projects.

2:00-3:00 Project Team Meeting
Discussions may be technical or administrative in nature. For my medical writing responsibilities, my main role is often to discuss the amount of time I will need to generate reports, in order to facilitate the discussion of project timelines. I also make the team aware of anything I will need to complete my tasks. For my project management responsibilities, my role is to make everyone aware of project timelines and budgets, to help identify and solve problems, and to facilitate communication between team members.

3:00-4:30 More follow-up and organization of projects.
Tasks include following up with appropriate team members about client and project needs, typing up meeting minutes, and working on my medical writing responsibilities. If I call a client for any reason, I take careful notes and type up a phone report, so that I have a written record of what was said and promised during the phone call. I also make appropriate team members aware of the conversation and what was said.

4:30-5:30 Work on internal administrative tasks for the projects.
Internal administrative tasks include reviewing the amount of time that each team member has billed to various aspects of the project. I also evaluate budgets in relation to the amount of work that still needs to be completed. Other administrative tasks include generating reports and graphs for senior management.

5:30-6:00 Work on internal, non-billable administrative tasks.
Tasks may include learning how to use software, reading journals, doing literature searches, and/or participating in other types of professional development or company sales (such as presentations to prospective clients). I also must keep my supervisor informed of my workload for the next several months.

6:00-6:15 One last pass-through of e-mail and voice mail before going home.

Other tasks: Sometimes I have to prepare for and attend client meetings. These can be for existing projects, or they can be sales meetings for potential projects. I also must meet occasionally with my supervisor and with upper management (for my project management responsibilities). Sometimes I am asked to provide feedback to supervisors of project team members for their employee evaluations. Once a year I have the opportunity to attend a professional meeting.

My day is sometimes longer than the sample day that I have provided above, and sometimes it is shorter. Although my hours are officially 8:30-5:30 with a one-hour lunch, the company offers considerable flexibility in working hours.

Aspects of my job that I like: I enjoy the interaction with people as well as the scientific and technical nature of the work. I also like being a leader. I enjoy the diversity of projects that I can work on in a contract research organization.

Aspects of my job that I do not like: I sometimes do not like being in a service industry in which the client is always right, no matter what. I also do not like the concept of billable time because I like to focus on tasks rather than on the time it takes to complete them. Because of short timelines and lean budgets, I do not always become as familiar with certain projects as I would like.

Other comments: I enjoy working at my company. The job market in clinical research is good, and this job offers me the potential to be promoted to higher-level, higher-paying positions within the pharmaceutical and biotechnology industries. The pay is decent. Starting salaries are similar to those of assistant professors at universities.

I sometimes miss the creative and independent aspects of bench research. Overall, though, I am happy with my decision to change career directions. My scientific training is still utilized in various ways, for example, to understand the pathogenesis of disease as well as good study design and conduct, data management and manipulation, and the presentation and interpretation of results. I am working considerably shorter hours than I was in the laboratory. I enjoy the interaction with people that I have in my present position.

How to get a job like mine: I was hired into my job directly from a postdoctoral position in a research laboratory. My lack of experience in the pharmaceutical industry was somewhat of a hindrance in obtaining a position within the industry, but I was able to use my manuscript writing experience in order to sell my abilities as a writer. A Ph.D. is helpful for this type of position, particularly for promotions. A Pharm.D. is also a popular and promotable degree within this industry. A doctoral degree is not always essential, though; it is possible to be hired as a medical writer with a bachelors or a masters degree.

I first heard about this type of career from an alumna at my graduate school, and also from trade publications such as *Genetic Engineering News*. I identified position openings by calling people from directories of professional societies. I did not know most of the people whom I called, but I called people who had position titles which I thought might be interesting. I asked them about their jobs and whether they had any tips for getting hired into their type of position. People were often very helpful, and I was grateful for that. An alumna from my graduate school was also able and willing to advise me and to help me identify reputable companies and job openings. She was a tremendous help to me.

It is unusual to have the opportunity to do project management this early in a person's career. Project management usually requires several years of experience in the industry. Because of the way my company operates, I was able to do it after only 8 months of experience.

A Day in the Life of:

Dolphin Dave, B.S.
Visitor Education Manager
An Aquarium

Description of Employer and Position: I work for a non-profit institution with a worldwide collection of habitats and animals housed in two buildings and over 2 million gallons of water. The largest displays feature an Atlantic coral reef, open ocean sharks, South American rain forest, rays, and bottlenose dolphins. Over 1.5 million people visit the aquarium each year, and we hope that the displays and presentations are educational and fun and inspire conservation stewardship of habitats and species.

As a Visitor Education Manager, I am responsible for two areas that impact our general visitor. First, I work in teams to develop exhibits and animal presentations. My part is to research, write, and evaluate the exhibits and scripts. I try to take the most current scientific research and translate it into clear and concise information for all ages of visitors. For our visitors that extend beyond our walls, we also have a website, and we answer information requests by mail. Second, I coordinate special events and educational programming that visitors can join, such as an open classroom of activities or auditorium presentations. This means not only managing staff but also working the programs myself.

Daily Activities: I check e-mail, voice mail, internal routings, and U.S. mail when I arrive at work in the morning. Mornings are my most productive time and I try to reserve blocks to do my own research or projects. Very often I have one meeting; most of mine are interdepartmental rather than outside people. During the summer I may also do some programming for general visitors in the morning or early afternoon.

We have a very flexible schedule, so my 45-minute lunch is anytime; or I may skip it altogether when I'm too busy.

I try to schedule meetings in the afternoon when I'm less productive and have the morning if I need to prepare. I have regular project meetings as well as brainstorming or staff meetings.

Before leaving, I try to outline any meeting notes, return and make phone calls, and check e-mail.

My day depends on the season and projects I have - summers are filled with staff management and program scheduling, winter/spring are always deadlines for a new exhibit or show, and the fall is my catch-up and planning period.

Other tasks: I overlap with some other program areas on occasion; I may do school programs, member programs, volunteer or teacher training throughout the year. I also work with other departments, from developing institutional programs to proofing written material for scientific accuracy. Some of my projects require new skills, such as working on computer interactive programs or our website.

Finally, I also try to keep professional development in mind. I generally present at 1-2 conferences a year and belong to different groups, which gives me a supportive network of colleagues in marine education.

Aspects of my job that I like: The aquarium has a diverse collection, and I like working on one area of interpretation and then moving on to another. I juggle several projects at once, so I tend not to get bored. For instance, it takes several months to develop and produce a dolphin show, and then rehearse it. I may be sick of dolphins after that but may also be halfway through a new exhibit on a very different topic or thinking about summer programs for visitors or compiling evaluation results from another project. All of my projects combine my science background with a lot of creativity, which is challenging and fun.

I have a lot of autonomy that goes with responsibility and often get to plan my own projects within the aquarium's scope. It's great to work within an organization with name recognition and with solid resources.

Aspects of my job that I do not like: We have a fairly large institution, and it takes patience to push projects through the bureaucracy. People associate my job with animals and with kids or teaching, but I spend more time working in project teams that include marketing, development, exhibit, and husbandry staff. Meetings can get tedious.

Also, we deal in worldwide habitats but are located in a busy part of a city. I wish we had the luxury of open space and shoreline to do field teaching. Instead, we are more like a museum and use our collection from within our walls. This is a common aspect of aquariums, because they are usually built to receive tourists in large cities.

Other comments: Because we are a public institution, a lot of our education programs and events are set for weekends or evening hours. I end up with a pretty flexible schedule but always put in a lot of long days or weeks in the summer, our busiest visitor time.

When I talk with students asking about marine careers, I always point out that marine education offers diversity and has the potential to impact public attitude and behavior toward marine resources at all age levels.

How to get a job like mine: I came into marine education at an entry level instructor position and have moved up within this institution. In college, I worked as a lab instructor and spent a summer as a YMCA camp counselor, which both helped, and I also did fieldwork in school. My degree is a BA in biology, but I had a fair number of marine/aquatic courses and a lot of English.

When I am hiring someone, I look at college transcripts (for near-entry-level applicants), experiences like internships in any informal setting, and fieldwork for authentic experiences. I also ask for writing samples to make sure a person is well-rounded and can communicate at all levels. Some institutions require a teaching certificate. We don't and are looking for a strong science background with the ability and enthusiasm to teach.

Starting salaries are now about high teens to low twenties for instructors or marine education specialists. Like most positions with informal education, they don't go very high. However, marine education is an expanding field; new aquariums are in construction now throughout the world. Zoos are also adding ocean exhibits, so that widens the field.

A good place to start looking for a job is to contact the American Association of Zoos and Aquariums (AZA) either through their website (http://www.aza.org) or by mail to: 7970-D Old Georgetown Road, Bethesda, MD 20814-2493. This group maintains a listing of all accredited institutions and has information about careers and current job opportunities and internships. Marine education is also an important part of government positions that deal with estuarine and coastal management.

Part V
Appendices:
Information and
Resources

Appendices

Table of Contents

Appendix A
Positions and Starting Salaries - Some Examples[1]

Type of Work	Position	Entry-Level Salary Range[2]	Educational Requirements[3]
Auditors, Inspectors, and Scientific Reviewers	Quality Control Inspector	$15,000 - $40,000	Bachelors or Masters
	Reviewer of Technical Applications	$35,000 - $60,000	Bachelors through Doctorate
	Industrial Hygienist	$18,000 - $35,000	Bachelors or Masters
Basic Scientific Research	Research Technician in Academia	$18,000 - $25,000	Associates through Masters
	Research Technician in Industry	$25,000 - $55,000	Associates through Masters
	Research Scientist in Academia (non-tenure-track)	$25,000 - $40,000	Doctorate
Clinical and/or Epidemiological Research	Clinical Research Study Monitor	$32,000 - $45,000	Bachelors through Doctorate
	Clinical Research Study Coordinator (at M.D. Investigator Sites)	$28,000 - $40,000	Bachelors through Doctorate
	Database Manager	$35,000 - $50,000	Bachelors or Masters
	Biostatistician	$40,000 - $55,000	Masters or Ph.D. in Statistics or Biostatistics
	Computer Programmer	$35,000 - $50,000	Bachelors in Science plus some computer training

Type of Work	Position	Entry-Level Salary Range[2]	Educational Requirements[3]
Clinical and/or Epidemiological Research (continued)	Regulatory Affairs Specialist	$35,000 - $55,000	Bachelors through Doctorate. Usually not entry-level position - some experience is often required. Some schools are now offering Masters degrees in Regulatory Affairs.
	Drug Safety Specialist	$32,000 - $55,000	Bachelors through Doctorate
	Medical Monitor for Clinical or Epidemiological Studies	$50,000 - $100,000	Doctorate (usually Medical Doctor (M.D.)
Laboratory Positions (other than those mentioned above)	Medical Technologist	$22,000 - $35,000	Bachelors or Masters plus Medical Technology certification
	Forensic Scientist	$25,000 - $50,000	Bachelors or Masters in Forensic Science, certification sometimes required
	Clinical Laboratory Director	$40,000 - $80,000	Doctorate plus postdoctoral clinical laboratory training
Law, Contracts, and Policy	Patent Lawyer	$40,000 - $90,000	Bachelors through Doctorate in Biological Sciences, Doctor of Jurisprudence (J.D.)

Type of Work	Position	Entry-Level Salary Range[2]	Educational Requirements[3]
Law, Contracts, and Policy (continued)	Legal Assistant	$30,000 - $40,000	Bachelors through Doctorate in Biological Sciences
	Technology Transfer Specialist	$35,000 - $60,000	Bachelors through Doctorate in Biological Sciences, Masters of Business Administration (M.B.A.) or Doctor of Jurisprudence (J.D.) sometimes desired as well
	Policy Analyst	$25,000 - $60,000	Masters or Doctorate
Management	Project, Program, or Grant Manager	$40,000 - $55,000	Bachelors through Doctorate
	Business Manager	$45,000 - $60,000	Bachelors through Doctorate in Biological Sciences plus Masters of Business Administration (M.B.A.)
Patient or Veterinary Care	Veterinarian	$28,000 - $55,000	Doctor of Veterinary Medicine (D.V.M.)
	Physician	$60,000 - $200,000	Medical Doctor (M.D.)
	Genetic Counselor	$32,000 - $42,000	Masters in Genetics, certification usually required
Sales	Product Salespeople	$32,000 - $45,000	Bachelors through Masters

Type of Work	Position	Entry-Level Salary Range[2]	Educational Requirements[3]
Scientific Writing	Medical Writer (e.g., for pharmaceutical/ biotechnology companies)	$35,000 - $55,000	Bachelors through Doctorate
	Scientific or Technical Writer (e.g., for news publications)	$20,000 - $55,000	Bachelors through Doctorate
Teaching	Assistant Professor (tenure track)	$25,000 - $55,000	Masters or Doctorate. Teaching certificate not usually required.
	Academic Instructor (non-tenure-track)	$20,000 - $45,000	Bachelors through Doctorate
	Teacher (elementary, junior high, or high school)	$15,000 - $30,000	Bachelors or masters. Teaching certificate usually required for public schools, and sometimes (but not always) for private schools.
Other	Technical Librarian	$25,000 - $40,000	Bachelors or higher in Biological Sciences, Masters of Library Science (M.L.S.)
	Interpreter for Museum, Aquarium, or Nature Preserve	$18,000 - $25,000	Bachelors through Doctorate
	Scientific Consultant (can be in virtually any scientific discipline, service, or topic area)	Wide Range	Bachelors through Doctorate

Appendix A
Footnotes

[1]By no means is Appendix A a comprehensive list of the careers that are available to you. These are just some examples. It is important to realize that many of these positions exist with many different types of employers. The job market for any specific field, as well as many other factors, will influence the education required and the salary offered for jobs of interest to you.

[2]Education requirements vary widely for positions depending on the employer, the job market in any specific industry and geographic region, and so on. These are just crude estimates. Professional societies, workers with similar jobs, and surveys are likely to contain more accurate information for your exact interests.

[3]Salary ranges vary widely for positions depending on the employer, your negotiating skills, your geographic area, the job market in any specific industry, and so on. These are just crude estimates and are not adjusted for geographic area or employer. Some positions may pay more than the examples given herein, and some may pay less, depending on many factors. Thus, you'll need to do your own homework to come up with realistic figures for your particular situation. Professional societies, workers with similar jobs, and surveys (such as those conducted by the U.S. Government Bureau of Labor Statistics), or by college or university career centers) are all good sources of information.

Please also consider the fact that salary is only part of your whole compensation package. When making career decisions, you'll also want to consider vacation time, benefits (for example, health, life, and disability insurance, sick leave, retirement benefits, and so on), work hours, commuting time and distance, and many other factors that can differ significantly from employer to employer.

Appendix B
Professional Societies

Professional Society	Contact Information
American College of Forensic Examiners	Central Administration 1658 S. Cobblestone Court Springfield, MO 65808 Phone: 417-881-3818 Fax: 417-881-4702 E-mail: acfedoc@aol.com
American Association for the Advancement of Science	1333 H Street, NW Washington, DC 20005 Phone: 202-326-6440 Fax: 202-789-0455 Website: http://www.aaas.org
American Medical Writers Association	9650 Rockville Pike Bethesda, MD 20814 Phone: 301-493-0003 Fax: 301-493-6384 E-mail: amwa@amwa.org
American Association of Pharmaceutical Scientists	1650 King Street Alexandria, VA 22314-2747 Phone: 703-548-3000 Fax: 703-684-7349 E-mail: membership@aaps.org
American Academy of Forensic Scientists	P.O. Box 669 Colorado Springs, CO 80901-0669 Phone: 719-636-1100 Fax: 719-636-1993 E-mail: membship@aafs.org
American Chemical Society	1155 16th Street, NW Washington, DC 20036 Phone: 800-227-5558 or 202-872-4600 Website: http://www.acs.org
American Statistical Society	1429 Duke Street Alexandria, VA 22314-3402

Professional Society	Contact Information
American Society for Quality Control	P.O. Box 3005 Milwaukee, WI 53201-3005 Phone: 800-248-1946 or 414-272-8575 E-mail: asqc@asqc.org Website: http://www.asqc.org
American Society for Microbiology	1325 Massachusetts Ave., N.W. Washington, DC 20005-4171 Phone: 202-737-3600 E-mail: membership@asmusa.org Website: http://www.asmusa.org
American Society of Human Genetics	9650 Rockville Pike Bethesda, MD 20814-3998 Phone: 301-571-1825 Fax: 301-530-7079 E-mail: society@genetics.faseb.org
Associates of Clinical Research Professionals	1012 14th Street, N.W. Washington, DC 20005 Phone: 202-737-8100 Fax: 202-737-8101 Website: http://www.acrpnet.org
Biotechnology Industry Organization	1625 K Street, N.W. Suite 1100 Washington, DC 20006 Phone: 202-857-0244 Fax: 202-857-0237 Website: http://www.bio.org
Drug Information Association	321 Norristown Road Suite 225 Ambler, PA 19002-2755 Phone: 215-628-2288 Fax: 215-641-1229 Website: http://www.diahome.org
Federation of American Societies for Experimental Biology (FASEB)	9650 Rockville Pike Bethesda, MD 20814 Phone: 800-43-FASEB or 301-530-7020 Fax: 301-571-0699 Website: http://www.faseb.org

Professional Society	Contact Information
National Association of Science Writers	P.O. Box 294 Greenlawn, NY 11740 Phone: 516-757-5664 E-mail: diane@nasw.org Website: http://www.nasw.org
National Business Incubation Association	20 East Circle Drive Suite 190 Athens, OH 45701 Phone: 614-593-4331 Fax: 614-593-1996 Website: http://www.nbia.org
National Science Teachers Association	1840 Wilson Blvd. Arlington, VA 22201-3000 Phone: 800-830-3232 or 703-243-7100 Website: http://www.nsta.org
National Society for Internships and Experiential Education	3509 Haworth Drive Suite 107 Raleigh, NC 27609 Phone: 919-787-3263 Website: http://www.tripod.com\nsee
National Society of Genetic Counselors	233 Canterbury Drive Wallingford, PA 19086 Phone: 610-872-7608
Parenteral Drug Association	7500 Old Georgetown Road Suite 260 Bethesda, MD 20814 Phone: 301-986-0293 Fax: 301-986-0296 Website: http://www.pda.org
Regulatory Affairs Professionals Society	12300 Twinbrook Parkway Suite 350 Rockville, MD 20852 Phone: 301-770-2920 Fax: 301-770-2924 Website: http://www.raps.org

Professional Society	Contact Information
Society for Epidemiologic Research	c/o American Journal of Epidemiology 2007 East Monument Street Baltimore, MD 21205 Phone: 301-955-3441
Society of Clinical Research Associates	P.O. Box 699 Hudson, OH 44236-0699 Phone: 800-SoCRA92 Website: http://www.uchsc.edu/misc.scra
Society of Quality Assurance	515 King Street Suite 420 Alexandria, VA 22314 Phone: 703-684-4050 Fax: 703-684-6048
The National Association for Interpretation	P.O. Box 1892 Fort Collins, CO 80522 Phone: 970-484-8283 Fax: 970-484-8179 E-mail: naimbrmgr@aol.com

There is a professional society for virtually every scientific specialty, and it is beyond the scope of this book to list them all. You are therefore encouraged to consult a librarian, university professor, or worker in your field for more information on the trade and professional societies in the individual scientific disciplines. A listing of professional societies is also provided in the following directory:

National Trade and Professional Associations of the U.S.

**Columbia Books
1212 New York Avenue, N.W., #330
Washington, DC 20005-3920
Phone: 202-898-0662**

Your local or university library is likely to have a copy of this directory. You can also explore the following website:

http://www.lib.uwaterloo.ca/society/overview.html

which provides links to all the professional societies that have servers on the Internet.

Appendix C
Publications and Resources (Hardcopy)

General News Publications

Publication	Contact Information
Genetic Engineering News	GEN Publishing, Inc. Mary Ann Liebert, Inc. 2 Madison Avenue Larchmont, NY 10538 Phone: 800-M-LIEBERT or 914-834-3100 Fax: 914-834-3688 E-mail: liebert@pipeline.com Website: http://www.genengnews.com

Comments: Covers many different scientific issues from a global perspective, including jobs. Also writes up Internet sites of interest to scientists. Available for a nominal ($19.95) mailing fee.

Nature	Subscription Department P.O. Box 5055 Brentwood, TN 37024-8743 Phone: 800-524-0384 Fax: 615-377-0525 E-mail: subscriptions@natureny.com Website: http://www.america.nature.com

Comments: Technically oriented. Also has a series of publications in various scientific specialties, such as *Nature Genetics*, *Nature Structural Biology*, *Nature Medicine*, and *Nature Biotechnology*. These publications cover hot topics pertinent to each of these scientific areas. Price: >$50.00

Occupational Outlook Handbook	Bureau of Labor Statistics Sales Center P.O. Box 2145 Chicago, IL 60690 Phone: 312-353-1880 Website: http://stats.bls.gov.ocohome.htm

Comments: These publications have information on employment in many different career fields. Cost: $32 (paper cover), $38 (hard cover).

News Publications *(continued)*

Publication	Contact Information
Occupational Outlook Quarterly	Superintendent of Documents P.O. Box 371954 Pittsburgh, PA 15250-7954 Phone: 202-512-1800 Fax: 202-512-2250

Comments: This publication, produced by the Bureau of Labor Statistics of the U.S. Federal Government, provides employment information and survey results. Additionally, each issue has an index of all publications for the last 5 years, so that you can look up topics of interest to you. Cost: $9.50 for a one-year subscription.

Science	American Association for the Advancement of Science 1333 H Street, N.W. Washington, DC 20005 Phone: 202-326-6440 Fax: 202-789-0455 Website: http://science-mag.aaas.org/science/

Comments: Technically oriented, can give a good overview of hot scientific topics. Price: >$50.00

The Scientist	3600 Market Street Suite 450 Philadelphia, PA 19104-2645 Phone: 215-386-9601 Fax: 215-387-7542 Website: http://165.123.33.33

Comments: Covers many different topic areas of interest to scientists, including jobs and current scientific issues. Internet sites relevant to topics of articles are often provided. Available free of charge or for a $19.95 fee.

Directories

Publication	Contact Information
Arco Book of Intern- ships	Arco Books Macmillan General Reference A Simon and Schuster Macmillan Company 1633 Broadway New York, NY 10019
Barron's Profile of American Colleges	Barron's Educational Series, Inc. 250 Wireless Boulevard Hauppauge, NY 11788

Comments: This directory of American colleges, like the Peterson's Guides, lists pertinent information about schools including their e-mail addresses and websites.

BioSupplyNet Source Book	10 Skyline Drive Plainview, NY 11803-9568 Phone: 516-349-5595 Fax: 516-349-5598 E-mail: info@biosupplynet.com Website: http://www.biosupplynet.com

Comments: This book is a buyer's guide for research products. It can be used by the job seeker to identify different companies who sell products to serve the research community. Cost: Free.

The *Genetic Engineer- ing News* Biotechnology Internet Address Book	GEN Publishing, Inc. Mary Ann Liebert, Inc. 2 Madison Avenue Larchmont, NY 10538 Phone: 800-M-LIEBERT or 914-834-3100 Fax: 914-834-3688 E-mail: liebert@pipeline.com

Comments: This book is organized similarly to the Genetic Engineering Guides to Biotechnology Companies (see below) but only contains the Internet addresses of these companies. Cost is cheaper than the Guides.

Directories *(continued)*

Publication	Contact Information
The *Genetic Engineering News* Guides to Biotechnology Companies	GEN Publishing, Inc. Mary Ann Liebert, Inc. 2 Madison Avenue Larchmont, NY 10538 Phone: 800-M-LIEBERT or 914-834-3100 Fax: 914-834-3688 E-mail: liebert@pipeline.com

Comments: This book is a directory which contains 12 guides in one, of organizations such as biotechnology companies, contract research organizations, lawyers and law firms specializing in biotechnology, biotechnology recruiters, technology transfer centers, and more. Cost: $215.

The Peterson's Guides to Higher Education	P.O. Box 2123 Princeton, NJ 08543-2123 Phone: 800-338-3282 Fax: 609-243-9150 Website: http://www.petersons.com

Comments: These books, which are organized by topic area, are directories of schools which offer post-secondary educational programs. Available at many libraries.

The Physicians Desk Reference	P.O. Box 10689 Des Moines, IA 50336 Phone: 800-232-7379 Fax: 201-573-4956

Comments: This book is written for physicians but can be used by the job seeker for addresses of pharmaceutical companies who have marketed drugs. Available at many medical libraries and medical researchers' offices.

The Princeton Review Publications on Internships	Phone: 800-273-8439

Directories (continued)

Publication	Contact Information
State-by-State Biotech- nology Directory	Bureau of National Affairs 1231 25th Street, N.W. Washington, DC 20037

Comments: This directory is published by the North Carolina Biotechnology Center and provides information to scientists regarding biotechnology. It may be available at some libraries.

The United States Government Manual	Superintendent of Documents P.O. Box 371954 Pittsburgh, PA 15250-7954 Phone: 202-512-1800 Fax: 202-512-2250

Comments: This book is an excellent resource for understanding the organization and structure of the U.S. federal government. This publication lists the subdivisions of the government and their mission statements, and it provides addresses to contact for employment. Available at most university and public libraries.

The Work-at-Home- Sourcebook, by Lynie Arden	Live Oak Publications P.O. Box 2193 Boulder, CO 80306 Phone: 800-247-6553

Comments: Lists jobs that can be done from home, and companies which have these types of jobs. Although it is not science-specific, it may get you started in your thinking about scientific careers that can be done from home.

Drugs, Biologics, and Biotechnology

Publication	Contact Information
The *Applied Clinical Trials Journal*	Advanstar Communications P.O. Box 6115 Dulmuth, MN 55806-6115 Phone: 908-549-3000 Fax: 908-549-8927

Comments: Covers many issues pertinent to clinical trials of drugs and biological agents. Contract research organizations in the pharmaceutical and biotechnology industries advertise here. Relevant Internet sites are often presented. Cost: Free.

BioConferences International Publications and Videotapes	Mary Ann Liebert, Inc. 2 Madison Avenue Larchmont, NY 10538 Phone: 800-5-BIO-CON

Comments: BioConferences International produces a number of videotapes and other instructional materials on drug and biologics development. Because the price is high (>$500 for a set of tapes), perhaps your institution or a group of friends could share the tapes.

Expediting Drug and Biologics Development, by Steven Linberg	Parexel International Corporation 195 West Road Waltham, MA 02154 Phone: 800-937-7795 or 617-487-9900 Fax: 617-487-0525

Comments: This book gives the reader an overview of the drug and biologics development process. Price: $135. There are also other books in this series.

New Drug Development: A Regulatory Overview, by Mark Matheiu	Parexel International Corporation 195 West Road Waltham, MA 02154 Phone: 800-937-7795 or 617-487-9900 Fax: 617-487-0525

Comments: This book gives the reader an overview of the drug development process from a regulatory standpoint. Price: $135. There are also other books in this series.

Drugs, Biologics, and Biotechnology (continued)

Publication	Contact Information
The North Carolina Biotechnology Center Publications	Biotechnology Information Division P.O. Box 14569 Research Triangle Park, NC 27709-4569 Phone: 919-544-5111 Fax: 919-544-5401

Comments: This organization publishes a directory of biotechnology companies. It also provides other information to scientists regarding biotechnology.

Teaching

Publication	Contact Information
Barron's Profile of American Colleges	Barron's Educational Series, Inc. 250 Wireless Boulevard Hauppauge, NY 11788

Comments: This directory of American colleges lists pertinent information about schools including their e-mail addresses and websites.

| *The Chronicle of Higher Education* | Subscription Department
The Chronicle of Higher Education
P.O. Box 1955
Marion, OH 43305
Phone: 800-728-2803
Fax: 202-223-7292
E-mail: circulation@chronicle.com
Website: http://chronicle.merit.edu |

Comments: Many teaching jobs are advertised here, including professorships at colleges. Cost: $75/year. Many libraries have it.

| The Peterson's Guides to Higher Education | P.O. Box 2123
Princeton, NJ 08543-2123
Phone: 800-338-3282
Fax: 609-243-9150
Website: http://www.petersons.com |

Comments: These books, which are organized by topic area, are directories of schools which offer post-secondary educational programs. Available at many libraries.

General Career Guidance

Publication	Contact Information
Outside the Ivory Tower, by Margaret Newhouse, Ph.D.	Office of Career Services Faculty of Arts and Sciences 54 Dunster Street Harvard University Cambridge, MA 02138 Phone: 617-495-2595

Comments: This book is written for graduate students and Ph.D. graduates who want to explore their career options outside of academia. While this book is not specifically written for scientists, it contains information about the processes of self-assessment, exploration, and preparation which will help you to land a satisfying "alternative" career.

A Ph.D. Is Not Enough, by Peter J. Feibelman	Addison Wesley Publishing Company Reading, MA 01867

Comments: This book emphasizes scientific survival skills in the areas of job interviewing, writing resumes, and preparing grant proposals, scientific research papers, and talks. Price: $12.95

To Boldly Go, by Peter S. Fiske	American Geophysical Union 2000 Florida Avenue, N.W. Washington, DC 20009 Website: http://www.agu.org/careerguide

Comments: This book is written specifically for scientists, although it is not specific for biological sciences. Nonetheless, it contains helpful job-seeking hints for scientists in general.

What Color Is Your Parachute, by Richard Nelson Bolles	Ten Speed Press P.O. Box 7123 Berkeley, CA 94707

Comments: This book is a good overall career book, which contains many helpful hints about how to find a position in an area you'll like. Although this book is not specifically written for biological scientists, it is an excellent general resource.

Internet Books

Publication	Contact Information
The Guide to Internet Job Searching, by Margaret Riley, Frances Roehm, and Steve Oserman	The Public Library Association / American Library Association VGM Career Horizons NTC Publishing Group Chicago, IL

Comments: This book contains helpful information about how to use the Internet for your job search. It also contains references to specific Internet sites which may be of interest to you.

The Internet for Dummies, by John R. Levine and Carol Baroudi	IDG Books Worldwide, Inc. 155 Bovet Road Suite 310 San Mateo, CA 94402 Phone: 800-434-3422

Comments: As its title suggests, this book gives an overview of the tools and applications of the Internet.

Appendix D
Internet Resources

Online Guides and Directories

A Biologist's Guide to Internet Resources, by Una Smith
There are several ways to access this site:

• anonymous ftp to **sunsite.unc.edu**
then go to
pub/academic/biology/ecology+evolution/bioguide/bioguide.faq

• gopher to **sunsite.unc.edu**
then select
"worlds of sunSITE - by Subject"
then select
"Ecology and Evolution"
then select
"A Biologist's Guide"

• E-mail to **ftpmail@sunsite.unc.edu**
with the following message:
open
cdpub/academic/biology/ecology+evolution/bioguide
get bioguide.faq
get README
quit

Comments: This guide provides a general overview of the different tools of the Internet. It also provides information on Internet sites such as Usenet groups and mailing lists which are of interest to biologists in different topic areas. The guide is updated periodically and can be obtained online free of charge. Highly recommended!

Sprynet
http://www.spry.com

Comments: This web site provides an overview of the different tools and applications of the Internet.

New Clearinghouse URL
http://www.lib.umich.edu/chhome.html

Comments: This site provides a clearinghouse of other subject-oriented Internet resource guides.

Search Engines

Category	Internet Address
Lists of Search Engines	**http://home.netscape.com/home/internet-search.html** (Netscape Net Search)
	http://www.submit-it.com (Submit It!)
Some Popular Web Search Engines	**http://www.altavista.digital.com** (Alta Vista: Main Page)
	http://www.lycos.com (Welcome to Lycos)
	http://yahoo.com (Yahoo)
	http://webcrawler.com (WebCrawler Search)
	http://www.excite.com (Excite)
Some Search Engines Devoted Specifically to Job Seeking (non-biology-specific)	**http://infoseek.com** (Infoseek)
	http://www.occ.com (Online Career Center)
	http://www.espan.com (Welcome to E.span)
	http://www.monster.com (The Monster Board)

Biologically- and Scientifically-Oriented Websites

Web Address	Comments
http://www.bio.com	This site, "Bio Online: Life on the 'Net", contains links to a great deal of information to biologists including companies, tips on job searching, employment opportunities, a job search firm, and other interesting tidbits of information.
http://www.sciweb.com	This site, "SciWeb - The Life Science Home Page", provides information and news in the biotechnology and pharmaceutical sciences. It also has employment listings.
http://www.biospace.com	This site, "BioSpace - the Hub Site for Biotechnology", allows you to do keyword searches and search for information by company.
http://www.os.dhhs.gov	This site, the home page of the U.S. Department of Health and Human Services, provides information on Internet sites which are likely to be of interest to biologists. It also provides links to employment opportunities within academia and the private sector, as well as a connection to the federal government's Office of Personnel Management Electronic Bulletin Board, which lists all federal position openings.
http://www.megalinx.net:80/	This site, "MegaLinx Communications", provides an enormous number of links to sites which contain information on employment opportunities, venture firms, bioconsultants, laboratory and other services, publishers, and other information of interest to biologists.
http://BioMedNet.com	This site, "BioMedNet", provides a listing of employment opportunities, as well as information and links to companies and products of interest to biological scientists.

Biologically- and Scientifically-Oriented Websites (continued)

Web Address	Comments
http://www.cato.com/interweb/cato/biotech	This site, the "World Wide Web Virtual Library: Biotechnology," provides directories of biotechnology firms and links to academic position openings as well as to other Internet sites and sources of information.
http://www.thomson.com/rcenters.biology/biology.html	This site, the "Wadsworth Biology Resource Center," has a career center with information on science careers and employment outlooks, as well as information on newsgroups and Internet sites of interest to biologists.
http://www.medmarket.com	This site, the "MED Market Virtual Industrial Park", has information on employment agencies and search firms, as well as some other interesting information for biological scientists.

Biologically- and Scientifically-Oriented Websites (continued)

Web Address	Comments
http://www.medsearch.com	This site, "MedSearch - Healthcare Careers," allows you to search for job openings by category and keyword. Although this site sounds like it is designed mainly for people who are seeking strictly medical employment, it includes basic scientist-type positions in academic laboratories and in biotechnology and pharmaceutical companies as well.
http://sci.aaas.org/nextwave	This site of the American Association for the Advancement of Science (AAAS), is an open forum for scientific discussions including role models who have gone into alternative science careers. It also has fellowship opportunities and discussions about careers in science in general (not biology-specific). It provides links to other Internet sites such as the AAAS home page, which lists career opportunities such as those that would be listed in the journal *Science*.
http://www.nsf.gov	This site is the National Science Foundation (NSF) home page. The NSF produces publications on scientific employment and other science issues.
http://www.nas.edu	This site of the National Academy of Sciences contains all sorts of information and hypertext links including news publications, reports, and the "Executive Summary of Reshaping the Graduate Education of Scientists and Engineers."
http://lcweb.loc.gov	This site of the Library of Congress provides access to all other federal agencies on the Internet.

Biologically- and Scientifically-Oriented Websites (continued)

Internet Address	Comments
http://www.physics.uiuc.edu/ysn **physics.uiuc.edu** (anonymous ftp) **ysnadm@crow-t-robot.stanford.edu** (e-mail for general administrative questions) **ysn-request@crow-t-robot-stanford.edu** (e-mail with the message "help" for information) **ysn-joblist@atlas.chemistry.iakron.edu** (e-mail with the message "send" for job listings) **ysnadmn@crow-t-robot-stanford.edu** (to subscribe to the YSN electronic newsletter, e-mail with the message "subscribe yourfirstname yourlastname")	The Young Scientists Network is a group of individuals who are committed, among other things, to discussing the employment situation among young scientists in both traditional and non-traditional careers.

Usenet Newsgroups

Internet Address	Comments
ftp://ftp.uu.net (login: anonymous password: your e-mail address) then type in: **cd usenet/news.answers.active-newsgroups** **get the files part1.Z and part2.Z**	This site provides a list of all active newsgroups. However, for biologically- and biomedically-oriented newgroups, you may well consult *A Biologist's Guide to Internet Resources* (please see previous pages for reference information).
bit.listserv.biojobs	This site is a newgroup which lists positions available in the Biojobs mailing list.
bionet.jobs.offered	This site is a newsgroup which lists positions available in the biological sciences.
sci.research.careers	This site is a newsgroup for discussions about careers in the sciences. It also has job listings.
sci.research.postdoc	This newsgroup is like the sci.research.careers newsgroup, except that it is designed specifically for discussions among postdoctoral students. It also has job listings.

Some Gopher Sites

**gopher://gopher.
micro.umn.edu:70**
  (select Other Gopher and Information Servers)
or go to
**gopher://gopher.
tc.umn.edu/11/Other%20Gopher%20and%20Information%20Servers/all**

<u>Comments:</u> This site is the central registry of all the gophers in the world.

gopher://gopher.gdb.org
  (select Search Databases at Hopkins)

<u>Comments:</u> This site provides a list of all Bio-oriented Internet gophers.

gopher://una.hh.lib.umich.edu/11/inetdirs/

<u>Comments:</u> This site, the New Clearinghouse URL, provides a clearinghouse of other subject-oriented Internet resource guides.

gopher://gopher.scs.unr.edu:70
  (select Search ALL of Gopherspace)

<u>Comments:</u> This Veronica site allows you to search for gopher sites by keyword.

gopher://riceinfo.rice.edu:70
  (select Information by Subject)

<u>Comments:</u> This site allows you to find gopher information by subject.

gopher://arl.cni.org
  (select Scholarly Communications, then Directory of Electronic Journals)

<u>Comments:</u> This site of the Association of Research Libraries contains a directory of electronic mailing lists, journals, and newsletters.

gopher://wcni.cis.umn.edu:11111

<u>Comments:</u> This site is the Academic Position Network.

Appendix E
List of Abbreviations

BA	Bachelor of Arts
BS	Bachelor of Science
CDC	Centers for Disease Control
CRO	Contract Research Organization
CV	Curriculum Vitae
DBU	Domestic Business Unit
DNA	Deoxyribonucleic Acid
DOE	Department of Energy
D.V.M.	Doctor of Veterinary Medicine
ELISA	Enzyme-Linked Immunosorbent Assay
EPA	Environmental Protection Agency
FBI	Federal Bureau of Investigation
FDA	Food and Drug Administration
FTP	File Transfer Protocol
HMO	Health Maintenance Organization
HPLC	High Performance Liquid Chromatography
IND	Investigational New Drug Application
IRS	Internal Revenue Service
J.D.	Doctor of Jurisprudence
M.B.A.	Masters of Business Administration
M.D.	Medical Doctor

List of Abbreviations
(continued)

M.L.S.	Master of Library Science
M.S.	Master of Science
NDA	New Drug Application
NIH	National Institutes of Health
OSHA	Occupational Safety and Health Administration
Pharm.D.	Doctor of Pharmacy
Ph.D.	Doctor of Philosophy
PLA	Product License Application
SOP	Standard Operating Procedure
URL	Uniform Resource Locator
U.S.	United States
USDA	United States Department of Agriculture
USUHS	Unformed Services University of the Health Sciences
WWW	World Wide Web

Index

Don't Risk Losing *This Book!*
Instead, Have Your
Friends
Order Their
Own Copies!

•Don't you **hate it** when books disappear into a black hole?

•Doesn't that **always** seem to **happen** when you lend books to your friends?

Prepare yourself for this inevitable event by photocopying this ordering information and handing the copies to your friends! You'll be helping them with their career planning while retaining your own copy of this book!

Ordering Information

Phone Orders: (800) 247-6553

Fax Orders: (419) 281-6883

Mail Orders: BookMasters, Inc.
                   P.O. Box 388
                   Ashland, OH 44805

E-Mail Orders: order@bookmaster.com
Visa, MasterCard, and personal checks accepted.

Price: **$22.95** ($19.95 + $3.00 S/H)   Foreign orders $29.95.

Please supply your name and shipping address!

Bulk discounts: Please contact Peer Productions, P.O. Box 13724, Research Triangle Park, NC 27709 for more information

Fax and Mail Order Form

Yes! I want to order a copy of "**Jump Start Your Career in BioScience**" by Chandra B. Louise, Ph.D.

Name: _____

Institution: _____

Position: _____

Mailing Address: _____

_____

_____

_____

Phone: _____

Fax: _____

E-Mail Address: _____

I have:

☐ Enclosed a check (made payable to BookMasters, Inc.)

☐ Included my credit card number:

_____ Visa or

Card Expiration Date: _____ MasterCard

Signature: _____ (please circle)

Price: $22.95 ($19.95 + $3.00 S/H) Foreign Orders $29.95

Please fax this form to:
BookMasters, Inc.
(419) 281-6883

or send it via U.S. mail to:
BookMasters, Inc.
P.O. Box 388
Ashland, OH 44805

Please note: You may want to photocopy this page and keep the original in the book!

Don't Risk Losing *This Book!*

Instead, Have Your
Friends
Order Their
Own Copies!

• Don't you **hate it** when books disappear into a black hole?

• Doesn't that **always** seem to **happen** when you lend books to your friends?

Prepare yourself for this inevitable event by photocopying this ordering information and handing the copies to your friends! You'll be helping them with their career planning while retaining your own copy of this book!

Ordering Information

Phone Orders: (800) 247-6553

Fax Orders: (419) 281-6883

Mail Orders: BookMasters, Inc.
P.O. Box 388
Ashland, OH 44805

E-Mail Orders: order@bookmaster.com
Visa, MasterCard, and personal checks accepted.

Price: **$22.95** ($19.95 + $3.00 S/H)   Foreign orders $29.95.

Please supply your name and shipping address!

Bulk discounts: Please contact Peer Productions, P.O. Box 13724, Research Triangle Park, NC 27709 for more information

Fax and Mail Order Form

Yes! I want to order a copy of "**Jump Start Your Career in BioScience**" by Chandra B. Louise, Ph.D.

Name: _____

Institution: _____

Position: _____

Mailing Address: _____

_____

_____

_____

Phone: _____

Fax: _____

E-Mail Address: _____

I have:

☐ Enclosed a check (made payable to BookMasters, Inc.)

☐ Included my credit card number:

_____

Visa or

Card Expiration Date: _____   MasterCard

Signature: _____   (please circle)

Price: $22.95 ($19.95 + $3.00 S/H) Foreign Orders $29.95

Please fax this form to:
BookMasters, Inc.
(419) 281-6883

or send it via U.S. mail to:
BookMasters, Inc.
P.O. Box 388
Ashland, OH 44805

Please note: You may want to photocopy this page and keep the original in the book!

Do You
Have
Comments
on How to
Improve This Book?

•Are there topics that you wish were covered in this book but weren't?

•Do you know of additional resources which should be mentioned in this book?

•Do you have any comments at all you wish to communicate with the author of this book?

The purpose of this book is to help biological scientists to find careers that suit their interests and needs. As such, the author of this book wants to know your feedback about how to improve this book to better serve you.

If you do have comments, please send them by U.S. Mail to:

Peer Productions
P.O. Box 13724
Research Triangle Park, NC 27709

or e-mail them to:

Peerpubs@aol.com

Thanks very much for your feedback!

Would **You** Benefit
from a
Live Presentation
by
Chandra Louise, Ph.D.?

Chandra B. Louise, Ph.D. can travel to your institution to speak about careers in the biosciences. She welcomes the opportunity to remain on-site after her seminar to meet with individual participants.

For further information about Chandra Louise's availability and rates, please photocopy this page and provide the following information:

Name of Your Institution: _____

Your Name: _____

Your Mailing Address: _____

_____

_____

_____

Phone: _____

Fax: _____

E-Mail Address: _____

Your Position within the Institution (professor, employee, student, postdoc, etc.): _____

Then send this completed form via U.S. Mail to:

Peer Productions
P.O. Box 13724
Research Triangle Park, NC 27709